FLOATING & SINKING

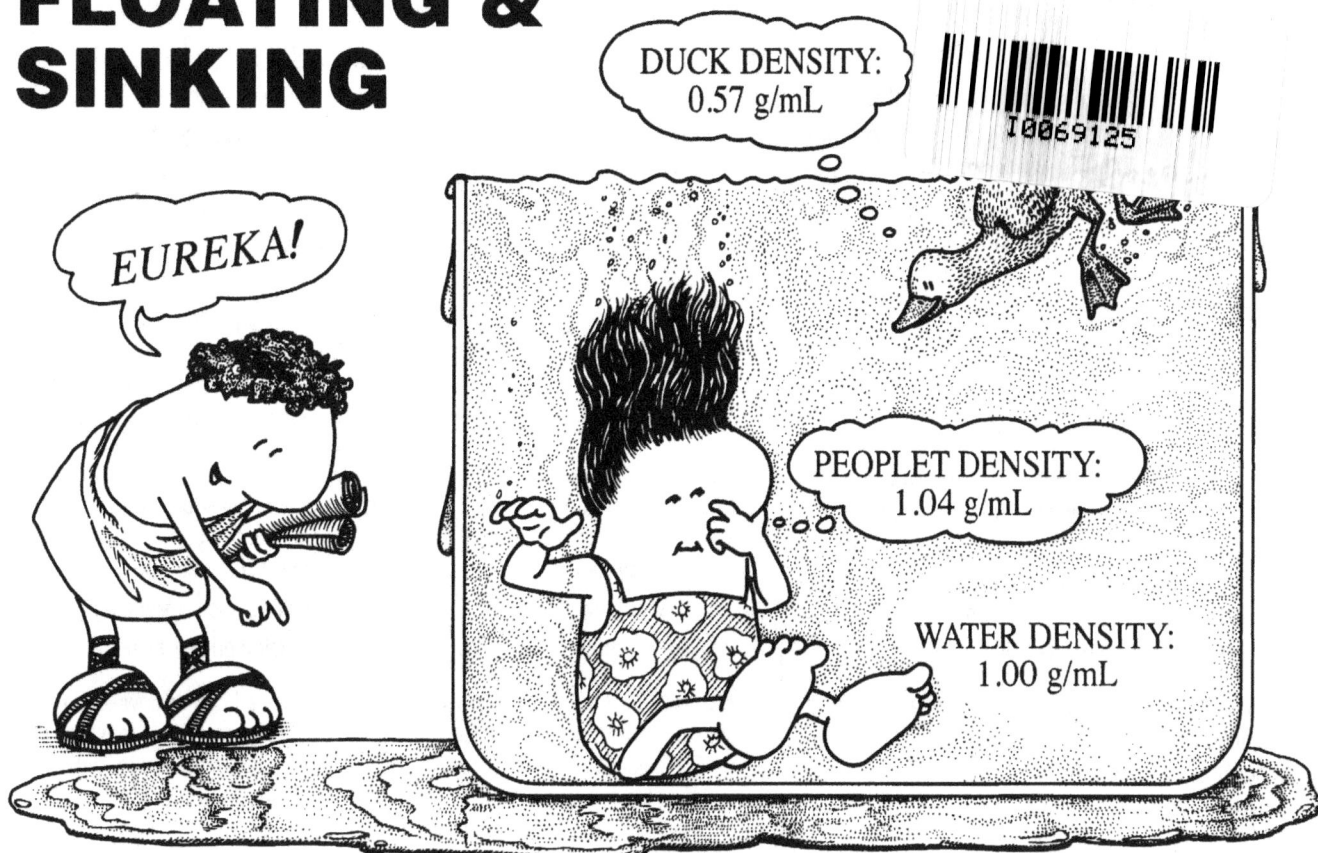

DUCK DENSITY:
0.57 g/mL

PEOPLET DENSITY:
1.04 g/mL

WATER DENSITY:
1.00 g/mL

EUREKA!

TASK CARD SERIES

Conceived and written by

RON MARSON

Illustrated by

PEG MARSON

**342 S Plumas Street
Willows, CA 95988**

www.topscience.org

TOPS LEARNING SYSTEMS

WHAT CAN YOU COPY?

Dear Educator,

Please honor our copyright restrictions. We offer liberal options and guidelines below with the intention of balancing your needs with ours. When you buy these labs and use them for your own teaching, you sustain our work. If you "loan" or circulate copies to others without compensating TOPS, you squeeze us financially, and make it harder for our small non-profit to survive. Our well-being rests in your hands. Please help us keep our low-cost, creative lessons available to students everywhere. Thank you!

PURCHASE, ROYALTY and LICENSE OPTIONS

TEACHERS, HOMESCHOOLERS, LIBRARIES:

We do all we can to keep our prices low. Like any business, we have ongoing expenses to meet. We trust our users to observe the terms of our copyright restrictions. While we prefer that all users purchase their own TOPS labs, we accept that real-life situations sometimes call for flexibility.

Reselling, trading, or loaning our materials is prohibited unless one or both parties contribute an Honor System Royalty as fair compensation for value received. We suggest the following amounts – let your conscience be your guide.

HONOR SYSTEM ROYALTIES: If making copies from a library, or sharing copies with colleagues, please calculate their value at 50 cents per lesson, or 25 cents for homeschoolers. This contribution may be made at our website or by mail (addresses at the bottom of this page). Any additional tax-deductible contributions to make our ongoing work possible will be accepted gratefully and used well.

Please follow through promptly on your good intentions. Stay legal, and do the right thing.

SCHOOLS, DISTRICTS, and HOMESCHOOL CO-OPS:

PURCHASE Option: Order a book in quantities equal to the number of target classrooms or homes, and receive quantity discounts. If you order 5 books or downloads, for example, then you have unrestricted use of this curriculum for any 5 classrooms or families per year for the life of your institution or co-op.

2-9 copies of any title: 90% of current catalog price + shipping.

10+ copies of any title: 80% of current catalog price + shipping.

ROYALTY/LICENSE Option: Purchase just one book or download *plus* photocopy or printing rights for a designated number of classrooms or families. If you pay for 5 additional Licenses, for example, then you have purchased reproduction rights for an entire book or download edition for any **6** classrooms or families per year for the life of your institution or co-op.

1-9 Licenses: 70% of current catalog price per designated classroom or home.

10+ Licenses: 60% of current catalog price per designated classroom or home.

WORKSHOPS and TEACHER TRAINING PROGRAMS:

We are grateful to all of you who spread the word about TOPS. Please limit copies to only those lessons you will be using, and collect all copyrighted materials afterward. No take-home copies, please. Copies of copies are strictly prohibited.

For licensing, honor system royalty payments, contact: **www.TOPScience.org**; or **TOPS Learning Systems 342 S Plumas St, Willows CA 95988**; or inquire at **customerservice@topscience.org**

ISBN 978-0-941008-79-2

CONTENTS

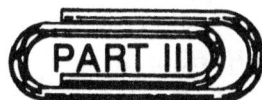

A TOPS Model for Effective Science Teaching...

If science were only a set of explanations and a collection of facts, you could teach it with blackboard and chalk. You could assign students to read chapters and answer the questions that followed. Good students would take notes, read the text, turn in assignments, then give you all this information back again on a final exam. Science is traditionally taught in this manner. Everybody learns the same body of information at the same time. Class togetherness is preserved.

But science is more than this.

Science is also process — a dynamic interaction of rational inquiry and creative play. Scientists probe, poke, handle, observe, question, think up theories, test ideas, jump to conclusions, make mistakes, revise, synthesize, communicate, disagree and discover. Students can understand science as process only if they are free to think and act like scientists, in a classroom that recognizes and honors individual differences.

Science is *both* a traditional body of knowledge *and* an individualized process of creative inquiry. Science as process cannot ignore tradition. We stand on the shoulders of those who have gone before. If each generation reinvents the wheel, there is no time to discover the stars. Nor can traditional science continue to evolve and redefine itself without process. Science without this cutting edge of discovery is a static, dead thing.

Here is a teaching model that combines the best of both elements into one integrated whole. It is only a model. Like any scientific theory, it must give way over time to new and better ideas. We challenge you to incorporate this TOPS model into your own teaching practice. Change it and make it better so it works for you.

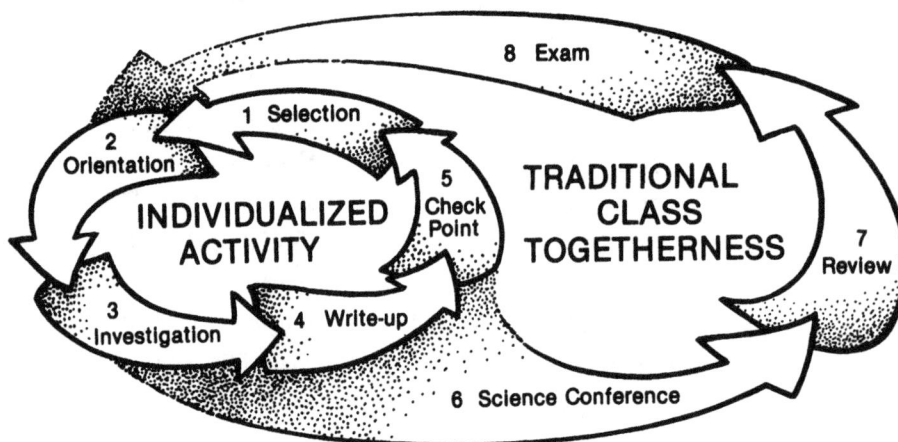

1. SELECTION

Doing TOPS is as easy as selecting the first task card and doing what it says, then the second, then the third, and so on. Working at their own pace, students fall into a natural routine that creates stability and order. They still have questions and problems, to be sure, but students know where they are and where they need to go.

Students generally select task cards in sequence because new concepts build on old ones in a specific order. There are, however, exceptions to this rule: students might *skip* a task that is not challenging; *repeat* a task with doubtful results; *add* a task of their own design to answer original "what would happen if" questions.

2. ORIENTATION

Many students will simply read a task card and immediately understand what to do. Others will require further verbal interpretation. Identify poor readers in your class. When they ask, "What does this mean?" they may be asking in reality, "Will you please read this card aloud?"

With such a diverse range of talent among students, how can you individualize activity and still hope to finish this module as a cohesive group? It's easy. By the time your most advanced students have completed all the task cards, including the enrichment series at the end, your slower students have at least completed the basic core curriculum. This core provides the common background so necessary for meaningful discussion, review and testing on a class basis.

3. INVESTIGATION

Students work through the task cards independently and cooperatively. They follow their own experimental strategies and help each other. You encourage this behavior by helping students only *after* they have tried to help themselves. As a resource person, you work to stay *out* of the center of attention, answering student questions rather than posing teacher questions.

When you need to speak to everyone at once, it is appropriate to interrupt individual task card activity and address the whole class, rather than repeat yourself over and over again. If you plan ahead, you'll find that most interruptions can fit into brief introductory remarks at the beginning of each new period.

4. WRITE-UP

Task cards ask students to explain the "how and why" of things. Write-ups are brief and to the point. Students may accelerate their pace through the task cards by writing these reports out of class.

Students may work alone or in cooperative lab groups. But each one must prepare an original write-up. These must be brought to the teacher for approval as soon as they are completed. Avoid dealing with too many write-ups near the end of the module, by enforcing this simple rule: each write-up must be approved *before* continuing on to the next task card.

5. CHECK POINT

The student and teacher evaluate each write-up together on a pass/no-pass basis. (Thus no time is wasted haggling over grades.) If the student has made reasonable effort consistent with individual ability, the write-up is checked off on a progress chart and included in the student's personal assignment folder or notebook kept on file in class.

Because the student is present when you evaluate, feedback is immediate and effective. A few seconds of this direct student-teacher interaction is surely more effective than 5 minutes worth of margin notes that students may or may not heed. Remember, you don't have to point out every error. Zero in on particulars. If reasonable effort has not been made, direct students to make specific improvements, and see you again for a follow-up check point.

A responsible lab assistant can double the amount of individual attention each student receives. If he or she is mature and respected by your students, have the assistant check the even-numbered write-ups while you check the odd ones. This will balance the work load and insure that all students receive equal treatment.

6. SCIENCE CONFERENCE

After individualized task card activity has ended, this is a time for students to come together, to discuss experimental results, to debate and draw conclusions. Slower students learn about the enrichment activities of faster students. Those who did original investigations, or made unusual discoveries, share this information with their peers, just like scientists at a real conference. This conference is open to films, newspaper articles and community speakers. It is a perfect time to consider the technological and social implications of the topic you are studying.

7. READ AND REVIEW

Does your school have an adopted science textbook? Do parts of your science syllabus still need to be covered? Now is the time to integrate other traditional science resources into your overall program. Your students already share a common background of hands-on lab work. With this shared base of experience, they can now read the text with greater understanding, think and problem-solve more successfully, communicate more effectively.

You might spend just a day on this step or an entire week. Finish with a review of key concepts in preparation for the final exam. Test questions in this module provide an excellent basis for discussion and study.

8. EXAM

Use any combination of the review/test questions, plus questions of your own, to determine how well students have mastered the concepts they've been learning. Those who finish your exam early might begin work on the first activity in the next new TOPS module.

Now that your class has completed a major TOPS learning cycle, it's time to start fresh with a brand new topic. Those who messed up and got behind don't need to stay there. Everyone begins the new topic on an equal footing. This frequent change of pace encourages your students to work hard, to enjoy what they learn, and thereby grow in scientific literacy.

GETTING READY

Here is a checklist of things to think about and preparations to make before your first lesson.

☐ Decide if this TOPS module is the best one to teach next.

TOPS modules are flexible. They can generally be scheduled in any order to meet your own class needs. Some lessons within certain modules, however, do require basic math skills or a knowledge of fundamental laboratory techniques. Review the task cards in this module now if you are not yet familiar with them. Decide whether you should teach any of these other TOPS modules first: *Measuring Length, Graphing, Metric Measure, Weighing* or *Electricity* (before *Magnetism*). It may be that your students already possess these requisite skills or that you can compensate with extra class discussion or special assistance.

☐ Number your task card masters in pencil.

The small number printed in the lower right corner of each task card shows its position within the overall series. If this ordering fits your schedule, copy each number into the blank parentheses directly above it at the top of the card. Be sure to use pencil rather than ink. You may decide to revise, upgrade or rearrange these task cards next time you teach this module. To do this, write your own better ideas on blank 4 x 6 index cards, and renumber them into the task card sequence wherever they fit best. In this manner, your curriculum will adapt and grow as you do.

☐ Copy your task card masters.

You have our permission to reproduce these task cards, for as long as you teach, with only 1 restriction: please limit the distribution of copies you make to the students you personally teach. Encourage other teachers who want to use this module to purchase their *own* copy. This supports TOPS financially, enabling us to continue publishing new TOPS modules for you. For a full list of task card options, please turn to the first task card masters numbered "cards 1-2."

☐ Collect needed materials.

Please see the opposite page.

☐ Organize a way to track completed assignment.

Keep write-ups on file in class. If you lack a vertical file, a box with a brick will serve. File folders or notebooks both make suitable assignment organizers. Students will feel a sense of accomplishment as they see their file folders grow heavy, or their notebooks fill up, with completed assignments. Easy reference and convenient review are assured, since all papers remain in one place.

Ask students to staple a sheet of numbered graph paper to the inside front cover of their file folder or notebook. Use this paper to track each student's progress through the module. Simply initial the corresponding task card number as students turn in each assignment.

☐ Review safety procedures.

Most TOPS experiments are safe even for small children. Certain lessons, however, require heat from a candle flame or Bunsen burner. Others require students to handle sharp objects like scissors, straight pins and razor blades. These task cards should not be attempted by immature students unless they are closely supervised. You might choose instead to turn these experiments into teacher demonstrations.

Unusual hazards are noted in the teaching notes and task cards where appropriate. But the curriculum cannot anticipate irresponsible behavior or negligence. It is ultimately the teacher's responsibility to see that common sense safety rules are followed at all times. Begin with these basic safety rules:

1. Eye Protection: Wear safety goggles when heating liquids or solids to high temperatures.
2. Poisons: Never taste anything unless told to do so.
3. Fire: Keep loose hair or clothing away from flames. Point test tubes which are heating away from your face and your neighbor's.
4. Glass Tubing: Don't force through stoppers. (The teacher should fit glass tubes to stoppers in advance, using a lubricant.)
5. Gas: Light the match first, before turning on the gas.

☐ Communicate your grading expectations.

Whatever your philosophy of grading, your students need to understand the standards you expect and how they will be assessed. Here is a grading scheme that counts individual effort, attitude and overall achievement. We think these 3 components deserve equal weight:

1. Pace (effort): Tally the number of check points you have initialed on the graph paper attached to each student's file folder or science notebook. Low ability students should be able to keep pace with gifted students, since write-ups are evaluated relative to individual performance standards. Students with absences or those who tend to work at a slow pace may (or may not) choose to overcome this disadvantage by assigning themselves more homework out of class.

2. Participation (attitude): This is a subjective grade assigned to reflect each student's attitude and class behavior. Active participators who work to capacity receive high marks. Inactive onlookers, who waste time in class and copy the results of others, receive low marks.

3. Exam (achievement): Task cards point toward generalizations that provide a base for hypothesizing and predicting. A final test over the entire module determines whether students understand relevant theory and can apply it in a predictive way.

Gathering Materials

Listed below is everything you'll need to teach this module. You already have many of these items. The rest are available from your supermarket, drugstore and hardware store. Laboratory supplies may be ordered through a science supply catalog.

Keep this classification key in mind as you review what's needed:

special in-a-box materials:	general on-the-shelf materials:
Italic type suggests that these materials are unusual. Keep these specialty items in a separate box. After you finish teaching this module, label the box for storage and put it away, ready to use again the next time you teach this module.	Normal type suggests that these materials are common. Keep these basics on shelves or in drawers that are readily accessible to your students. The next TOPS module you teach will likely utilize many of these same materials.
(substituted materials):	*optional materials:
Parentheses enclosing any item suggests a ready substitute. These alternatives may work just as well as the original, perhaps better. Don't be afraid to improvise, to make do with what you have.	An asterisk sets these items apart. They are nice to have, but you can easily live without them. They are probably not worth an extra trip to the store, unless you are gathering other materials as well.

Everything is listed in order of first use. Start gathering at the top of this list and work down. Ask students to bring recycled items from home. The teaching notes may occasionally suggest additional student activity under the heading "Extensions." Materials for these optional experiments are listed neither here nor in the teaching notes. Read the extension itself to find out what new materials, if any, are required.

Needed quantities depend on how many students you have, how you organize them into activity groups, and how you teach. Decide which of these 3 estimates best applies to you, then adjust quantities up or down as necessary:

$Q_1 / Q_2 / Q_3$

- **Single Student:** Enough for 1 student to do all the experiments.
- **Individualized Approach:** Enough for 30 students informally working in 10 lab groups, all self-paced.
- **Traditional Approach:** Enough for 30 students, organized into 10 lab groups, all doing the same lesson.

KEY:	*special in-a-box materials* *(substituted materials)*	general on-the-shelf materials *optional materials

1/10/10	small 10 mL graduated cylinders	1/3/3 toothpicks
1/1/1	source tap water, hot and cold	1/1/1 pkg birthday candles
1/10/10	gram balances – see notes 1	1/10/10 paper drinking cups, 6 oz. or larger
1/1/1	box table salt	1/5/10 medium-sized cans
2/2/2	quart jars or equivalent with lids	1/2/10 *golf balls*
1/1/1	bottle 70% isopropyl rubbing alcohol	2/10/20 size-D batteries, dead or alive
1/1/1	bottle 100% corn oil	1/4/10 centimeter rulers
1/1/1	bottle baby oil (mineral oil)	1/1/1 box plastic wrap
5/45/50	large 6 oz. baby food jars with lids	2/20/20 *natural corks – see notes 13*
5/45/50	eye droppers	1/10/10 plastic sandwich bags
1/1/1	roll paper towels	1/10/10 large cereal boxes – see notes 15
1/2/4	rolls masking tape	2/15/20 large rubber bands
1/5/10	jars or glasses, 8 oz. or more	1/1/1 *roll 28 ga steel wire, annealed or galvanized*
1/10/10	*test tubes (small graduates)	1/3/10 meter sticks
2/15/15	*utility candles – see notes 5*	1/10/10 paper clips
1/1/1	box straight pins	1/1/1 spool of thread
1/2/5	needle-nose pliers with wire cutting edge	1/10/10 *film canisters, plastic 35 mm size*
1/10/10	large 100 mL graduated cylinders	1/3/10 solid rubber stoppers
3/30/30	straight plastic drinking straws	1/3/10 books of matches
1/2/5	paper punchers	1/3/10 plastic lids
1/10/10	hand calculators	1/1/1 pkg BB shot
.5/5/5	cups oil-based modeling clay	1/3/10 tablespoons
1/10/10	scissors	1/1/1 bottle blue food coloring with dispenser
1/10/10	plastic margarine tubs (bowls)	1/10/10 ice cubes
1/10/10	*pieces wood dowel – see notes 8*	1/2/10 helium-filled balloons

D

Sequencing Task Cards

    This logic tree shows how all the task cards in this module tie together. In general, students begin at the bottom of the tree and work up through the related branches. As the diagram suggests, upper level activities build on lower level activities.

    At the teacher's discretion, certain activities can be omitted, or sequences changed, to meet specific class needs. The only activities that must be completed in sequence are indicated by leaves that open *vertically* into the ones above them. In these cases the lower activity is a prerequisite to the upper.

    When possible, students should complete the task cards in the same sequence as numbered. If time is short, however, or certain students need to catch up, you can use the logic tree to identify concept-related *horizontal* activities. Some of these might be omitted, since they serve only to reinforce learned concepts, rather than introduce new ones.

    On the other hand, if students complete all the activities at a certain horizontal concept level, then experience difficulty at the next higher level, you might move back down the logic tree to have students repeat specific key activities for greater reinforcement.

    For whatever reason, when you wish to make sequence changes, you'll find this logic tree a valuable reference. Parentheses in the upper right corner of each task card allow you total flexibility; they are left blank so you can pencil in sequence numbers of your own choosing.

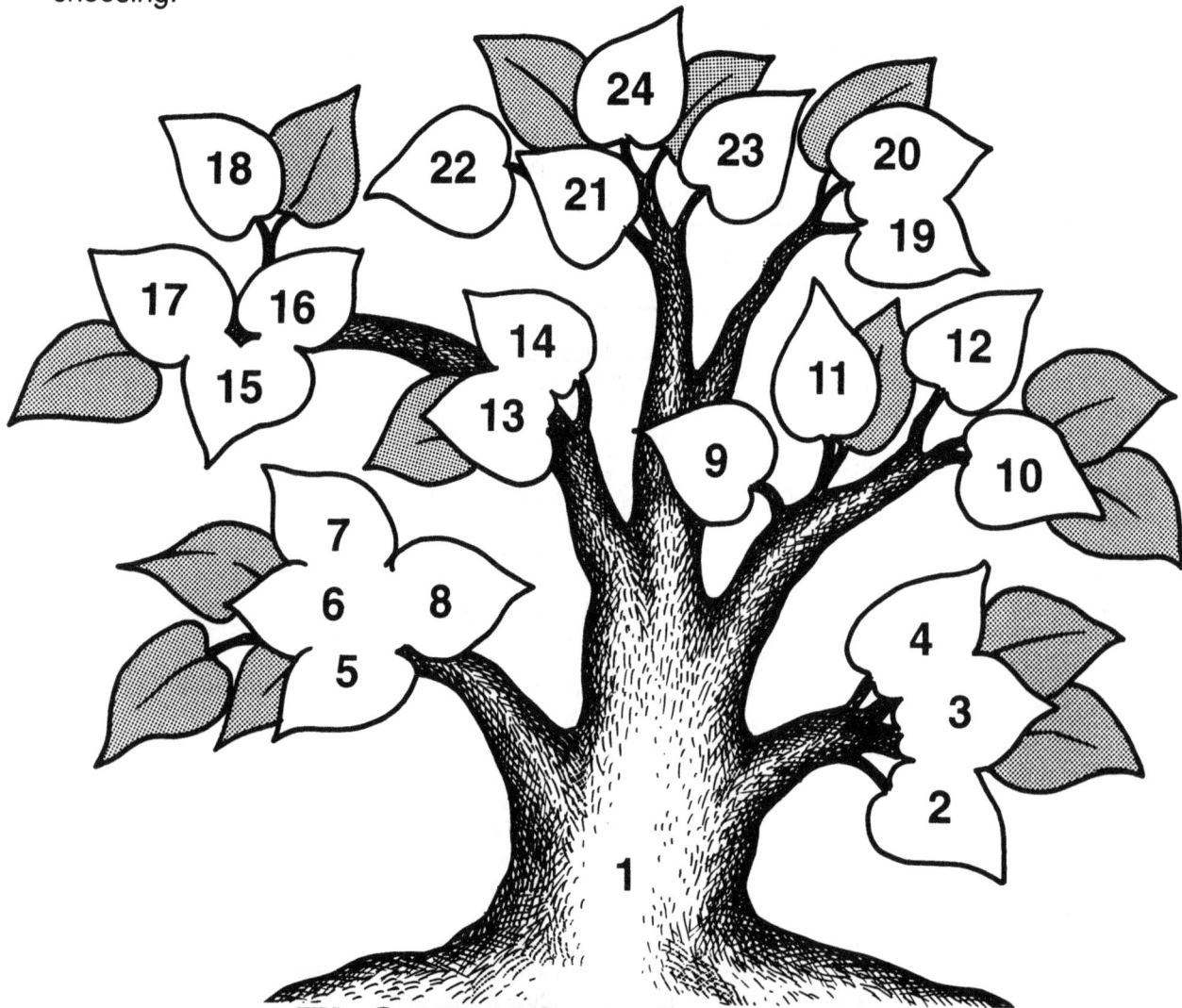

FLOATING & SINKING 09

LONG-RANGE OBJECTIVES

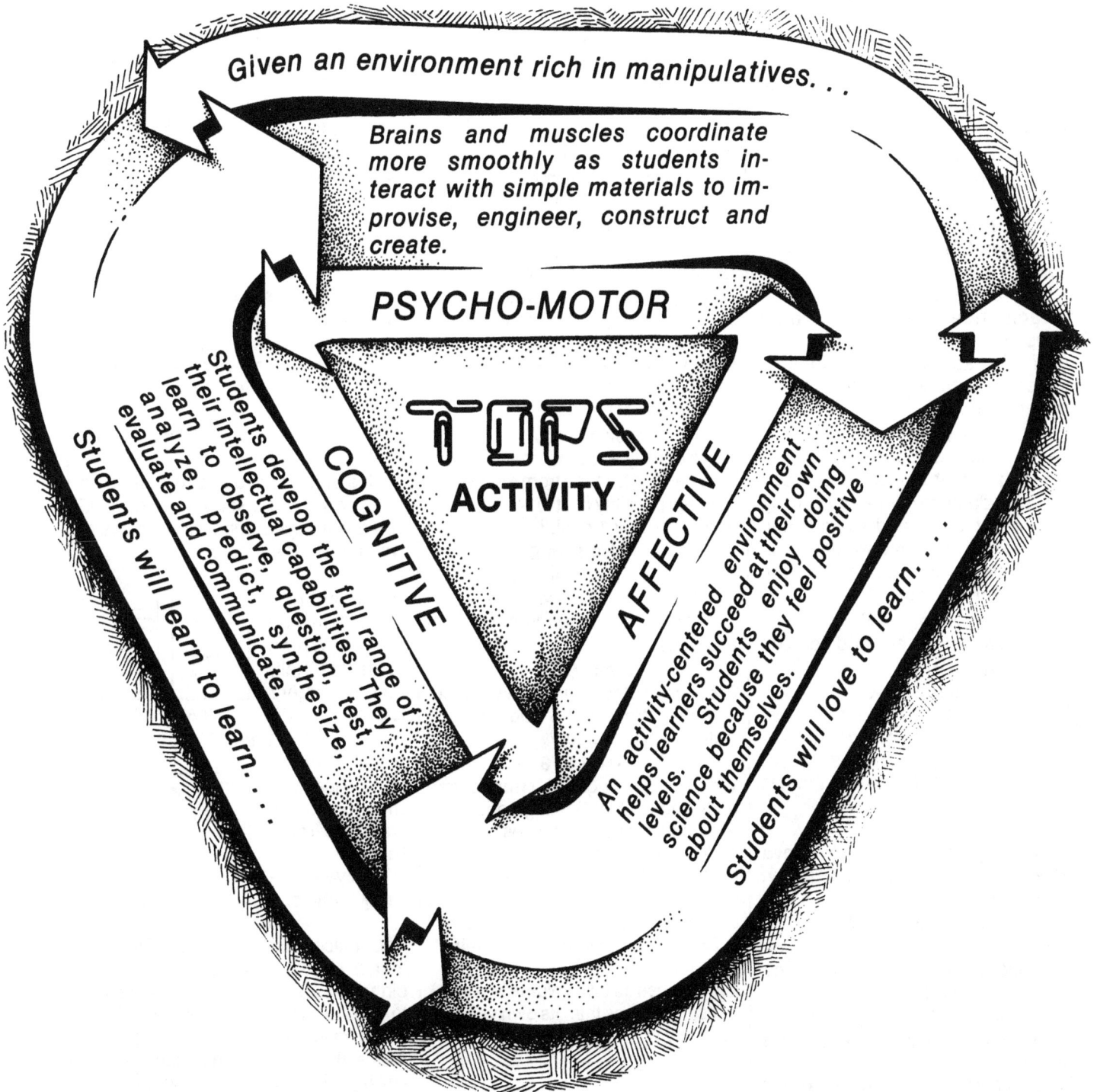

Given an environment rich in manipulatives. . .

Brains and muscles coordinate more smoothly as students interact with simple materials to improvise, engineer, construct and create.

PSYCHO-MOTOR

TOPS
ACTIVITY

COGNITIVE

Students develop the full range of their intellectual capabilities. They learn to observe, question, test, analyze, predict, synthesize, evaluate and communicate.

Students will learn to learn. . . .

AFFECTIVE

An activity-centered environment helps learners succeed at their own levels. Students enjoy doing science because they feel positive about themselves.

Students will love to learn. . . .

Review / Test Questions

task 1-2 A
A small graduated cylinder filled with 6.0 mL water has a mass of 34.7 g. What is the mass of the dry cylinder?

task 1-2 B
Is 10 g of water twice as dense as 20 g of water? Explain.

task 1-2 C
A 10.0 mL volume of kerosene has a mass of 8.2 g. What is the density of kerosene?

task 3-4 A
A block of wood (D = .84 g/mL) is immersed in a glass of kerosene (D = 0.82 g/mL). Will it float or sink? Explain.

task 3-4 B
Kerosene (D = 0.82 g/mL) and carbon tetrachloride (D = 1.56 g/mL) mix together. Neither of these liquids mixes with water. Diagram how a test tube looks:
a. If you add equal portions of carbon tetrachloride, then water, then kerosene.
b. If you add equal portions of kerosene, then water, then carbon tetrachloride.

task 5, 8-9 A
Hand calculator optional.
A 18.3 g rubber stopper is dropped into a graduated cylinder containing 50.0 mL of water and sinks to the bottom. This raises the water level in the cylinder to 65.4 mL. Calculate the density of the stopper.

task 5, 8-9 B
Design an experiment to determine the density of nails.

task 6, 8 A
A 10.0 cm candle sinks into a narrow water-filled cylinder, with only 0.8 cm remaining above the surface. Calculate its specific gravity. What is its density?

task 6, 8 B
An iceberg floats just 8.3% above water. Calculate its specific gravity. What is its density?

task 7, 8, 12, 14
A 25 ton iceberg floats in arctic waters. How much seawater does it displace?

task 4, 9
Solid A floats in liquids x, y and z.
Solid B floats in liquids x and y, but sinks in z.
Solid C floats in liquid x, but sinks in y and z.
  Arrange these 6 substances from lowest density to highest density.

task 10
A 30 mL portion of water is emptied into a box that measures 2 cm by 3 cm by 4 cm. Does the box overflow?

task 10-11
You'll need a hand calculator.
A sphere of granite has a radius of 2.0 cm and a density of 2.7 g/mL.
a. What is the volume of this perfectly round rock? (V = 4/3πr³)
b. What is its mass?

task 11
A 60.9 g mass of basalt has a density of 2.9 g/mL. What is the volume of this rock?

task 5, 7, 12, 14
An 85 g wood ball and a 715 g steel ball both have a volume of 100 mL. What volume of water will each ball displace? Explain your reasoning.

task 13
Several tablespoons of salt are poured into a tall glass of water without stirring. An egg placed in the glass sinks to the bottom and rests on the salt. Over several days, it rises halfway between the bottom and the surface. What happened?

task 12, 14 A
Iron's density is greater than 7 g/mL. How can boats made from this dense material possibly float in water with a density of only 1 g/mL?

task 12, 14 B
A swimmer tends to float when inhaling air, and sink when exhaling air. Explain this in terms of volume and density.

task 15-17
A 2.8 ton boulder weighs only 1.0 ton in water. What is the buoyancy of water on this boulder?

task 14-18
Write F for Floating body, N for Neutral body and/or S for Sinking body after each statement that applies:
a. Weighs nothing in water.
b. Weighs less in water than in air, but not zero.
c. Displaces its own weight in water.
d. Displaces less than its own weight in water.
e. Displaces its own volume in water.
f. Displaces less than its own volume in water.

task 17-18
Write F for Floating body or S for Sinking body after each statement that applies to fresh water (f.w.) and salt water (s.w.).
a. Weighs the same in f.w. as in s.w.
b. Weighs more in f.w. than in s.w.
c. Displaces less weight in f.w. than s.w.
d. Displaces same weight in f.w. as s.w.
e. Displaces higher volume f.w. than s.w.
f. Displaces same volume f.w. as s.w.

task 17-20
A ship sails out of the fresh water Columbia River into the salt water Pacific Ocean. Did it ride any higher or lower in the river than in the ocean? Use Archimedes' principle to defend your answer.

task 19-20 A
Explain how a hydrometer measures density.

task 19-20 B
You'll need a mm ruler.
Float levels for water (D = 1.00 g/mL) and rubbing alcohol (D= 0.87 g/mL) are marked on a hydrometer, with a separation of 13 mm. Draw a scale for this hydrometer. Use an arrow to show the float level for corn oil (D = 0.91 mL).

task 3, 4, 21
A blue salt is dissolved in unmarked bottles of alcohol and water. How would you use an eyedropper and a glass of water to tell them apart?

task 22
Which jar holds the half-melted wax? Which holds the half-melted water? Explain.

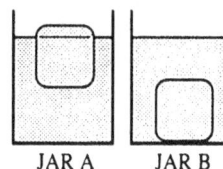

JAR A   JAR B

task 23
A nail is frozen into the ice cube that floats in water at room temperature. Will it eventually sink? Explain.

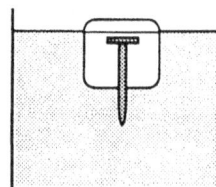

task 24
You notice, just before going to bed, that a balloon left from last week's birthday party has lost enough helium to sink to the floor of your bedroom. Upon rising the next morning, you notice that it has again risen to the ceiling. Did helium pixies recharge the balloon overnight? Offer a more plausible theory in terms of Archimedes' principle.

Answers

Photocopy the questions on the left. On a separate sheet of blank paper, cut and paste those questions you want to use in your test. Include questions of your own design, as well. Crowd all these questions onto a single page for students to answer on another paper, or leave space for student responses after each question, as you wish. Duplicate a class set, and your custom-made test is ready to use. Use leftover questions as a review in preparation for the final exam.

task 1-2 A
Mass of water = 6.0 g
Mass of dry cylinder = 34.7 g – 6.0 g
$$= 28.7 \text{ g}$$

task 1-2 B
No. The larger amount has twice the mass of the smaller and twice its volume. These divide to the same density:
10 g / 10 mL = 20 g / 20 mL = 1.0 g/mL

task 1-2 C
Density of kerosene =
8.2 g / 10.0 mL = 0.82 g/mL

task 3-4 A
The block of wood sinks in kerosene because it has a higher density.

task 3-4 B

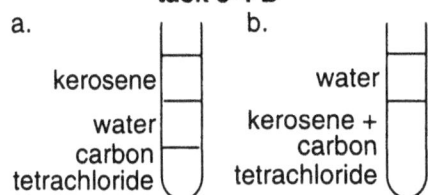

a.
kerosene
water
carbon tetrachloride

b.
water
kerosene + carbon tetrachloride

task 5, 8-9 A
Volume of stopper =
65.4 mL – 50.0 mL = 15.4 mL
Density of stopper =
18.3 g / 15.4 mL = 1.19 g/mL

task 5, 8-9 B
Find the mass of perhaps 10 nails on a gram balance. Put these into a graduated cylinder containing a measured quantity of water. Subtract the final water level from the initial level to determine the volume of water displaced by the nails. Divide mass by volume to calculate density.

task 6, 8 A
Length of candle under water =
10.0 cm – 0.8 cm = 9.2 cm.
S.G. = 9.2 cm / 10.0 cm = 0.92.
Density = S.G. = 0.92 g/mL

task 6, 8 B
Percent of iceberg below water =
100.0 % – 8.3 % = 91.7 %.
S.G. = 91.7 % / 100 % = .917
Density = S.G. = 0.917 g/mL

task 7, 8, 12, 14
The floating iceberg displaces 25 tons of seawater, equal to its own mass.

task 4, 9
(lowest density)   A, z, B, y, C, x   (highest density)

task 10
Yes. The box holds less than 30 mL:
2 cm x 3 cm x 4 cm = 24 cm³ = 24 mL

task 10-11
a. V = (4/3) (3.14) (2.0)³ =
33.5 cm³ = 33.5 mL.
b. Mass = 33.5 mL x 2.7 g/mL = 90.5 g

task 11
60.9 g x 1 / 2.9 g/mL = 21 mL

task 5, 7, 12, 14
The steel ball sinks in water, and thus displaces its own volume:
100 mL displaced water
The wood ball floats in water, and thus displaces is own mass:
85 g displaced water = 85 mL

task 13
Initially the egg is more dense than the water, resting on the bottom layer of salt. Over time, more and more salt dissolves into the water, increasing its density, especially near the bottom. This higher density salty water displaces the egg upward, into lower density, less salty water. There it floats, somewhere in the middle, at a level that equals the egg's own average density.

task 12, 14 A
The dense iron is rolled flat into sheets, and riveted into a boat shape that occupies a much greater volume than it did in its original form. This boat shape displaces its own mass before sinking too far into the water, since it now has an average density much lower than 1 g/mL.

task 12, 14 B
The swimmer increases body volume without significantly increasing mass by inhaling air, reducing density below that of water, and floating. Similarly, the swimmer decreases volume by exhaling, thereby increasing body density above that of water, and sinking.

task 15-17
Buoyancy of water =
2.8 tons - 1.0 ton = 1.2 tons.

task 14-18
a. F, N   b. S   c. F, N
d. S   e. N, S   f. F

task 17-18
a. F   b. S   c. S   d. F   e. F   f. S

task 17-20
The ship rides a bit higher in the salty Pacific. It displaces its own weight in both bodies of water. Since salt water is a little more dense, the ship displaces a little less of it to float.

task 19-20 A
It sinks lower in less dense fluids and rises higher in more dense fluids.

task 19-20 B

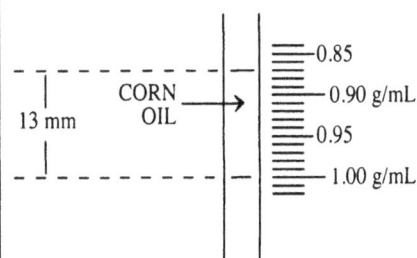

13 mm   CORN OIL →   0.85   0.90 g/mL   0.95   1.00 g/mL

task 3, 4, 21
Simply drip each blue salt solution into the glass of water. If the blue liquid sinks, it must be the higher density salt-water solution; if it remains at the surface it must be the lower density salt-alcohol solution.

task 22
Jar B holds the half-melted wax, which is more dense in its solid form, sinking in its own liquid. Jar A holds the half-melted ice, which is less dense in its solid form, floating in its own liquid.

task 23
Yes. The lower density ice will continue to melt while the higher density nail remains intact. Over time, the average density of both ice and nail will increase beyond the density of the surrounding water, causing both to sink.

task 24
Helium pixies were not involved. The balloon sank to the floor in the evening, because it lost enough volume to no longer displace its own weight in room air. Overnight, however, the air cooled and gained enough density to sink below the balloon and displace it up to the ceiling again.

H

TEACHING NOTES
For Activities 1-24

Task Objective (TO) experimentally confirm that the density of water is very close to 1.00 g/mL, no matter how much water you measure.

DENSITY OF WATER ◯   Floating and Sinking ( )

1. Get a 10 mL graduated cylinder and a gram balance. Center the balance.

a. Show that these *volumes* of water have the indicated *masses*, within the limits of measuring uncertainty:

b. Recenter your balance, as necessary, before finding each new mass.

c. Write a brief report.

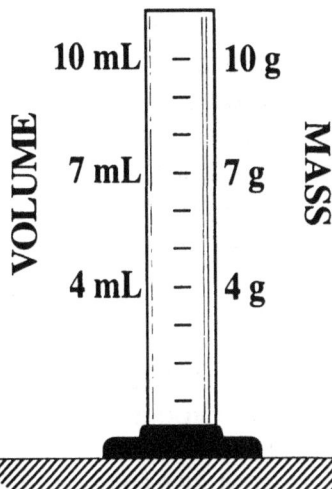

VOLUME		MASS
10 mL	–	10 g
7 mL	–	7 g
4 mL	–	4 g

2. *Density* is defined as the mass of any substance divided by its volume. Show that the density of water is always close to 1.00 g/mL, no matter what volume you measure.

$$D = \frac{mass}{volume} = \frac{g}{mL}$$

© 1995 by TOPS Learning Systems

Introduction

Review, if necessary, these experimental procedures:

a. How to read a graduated cylinder at the bottom of the meniscus.

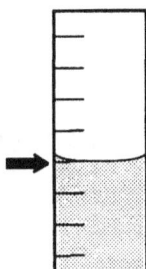

b. How to center a balance and measure mass in grams.

Materials

☐ A 10 mL capacity graduated cylinder. Larger 100 mL graduated cylinders do not measure volume with enough accuracy.

☐ A source of water. Water is used throughout this module and will hereafter be assumed.

☐ A gram balance. Use the equal-arm balance improvised in TOPS Module 05 Weighing, or any standard lab balance.

Answers / Notes

1. *Sample report for students using a triple-beam balance:*
Weigh a 10 mL graduate with the specified volume of water. Weigh the empty graduate. Find the difference.

	10.0 mL	7.0 mL	4.0 mL
volume of water	10.0 mL	7.0 mL	4.0 mL
mass of graduate + water	38.7 g	35.8 g	32.6 g
mass of empty graduate	28.7 g	28.7 g	28.7 g
mass of water	10.0 g	7.1 g	3.9 g

1. *Sample report for students using the TOPS equal-arm balance from 05 Weighing:*

First center the balance with a *wet*, empty cup. Press enough clay onto the clip over the dry cup to counterbalance the extra moisture, with the rubber band riders centered on each arm.

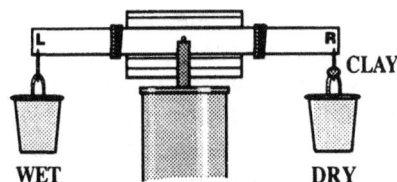

Pour a measured volume of water directly into the wet cup, then add gram masses (lighter ones made of paper, heavier ones made of rock and tape) to the dry cup until the beam rebalances at its centered position.

Empty the water cup and recenter the balance after each weighing. After the initial clay placement, do this by shifting the rubber band riders and/or tilting the index card to align with the beam.

volume of water (mL)	10 mL	7 mL	4 mL
mass of water	9.9 g	6.9 g	4.0 g

2. In every case the density of water is always the same:
D = 10.0 g/10.0 mL ≈ 7.1g/7.0 mL ≈ 3.9 g/4.0 mL ≈ 1.00 g/mL

(TO) experimentally determine the densities of 4 different liquids. This data will be used in later activities.

DENSITY OF OTHER LIQUIDS ◯   Floating and Sinking ( )

1. You already found that the density of water is 1.00 g/mL. Now calculate the density of each of these liquids, following the steps below:

   a. Rinse and dry your 10 mL cylinder between each new liquid. Roll and tape a paper towel to reach to the bottom.

SALT WATER,   RUBBING ALCOHOL,   CORN OIL,   BABY OIL

   b. Find the mass of 10.0 mL samples of each liquid to the nearest 0.1 g. Recenter your balance, if necessary, before finding the mass of each new liquid.

   c. Report all densities to the nearest 0.01 g/mL. Show your math.

2. Order these 4 liquids, plus water, from most dense to least dense.

MOST DENSE   ——————————▶   **LEAST DENSE**

© 1995 by TOPS Learning Systems   2

Answers / Notes

1. *After each mass determination, students should pour their 10 mL samples down the sink, not back into the dispensing containers.*

    Density salt water = 11.8 g / 10.0 mL = 1.18 g/mL
    Density rubbing alcohol = 8.7 g / 10.0 mL = 0.87 g/mL
    Density corn oil = 9.1 g / 10.0 mL = 0.91 g/mL
    Density baby oil = 8.3 g / 10.0 mL = 0.83 g/mL

2. MOST DENSE ——— salt water, fresh water, corn oil, rubbing alcohol, baby oil ➔ LEAST DENSE

Materials

☐ A 10 mL graduated cylinder.
☐ A gram balance.
☐ Saturated salt water: add 1 part table salt to 5 parts water in a sealed quart jar or other large container; shake vigorously for at least 1 minute; let settle overnight; pour off clear liquid into another clean jar. Save the undissolved residue to make more saturated solution in later activities.
☐ 70% isopropyl rubbing alcohol. Check the label; rubbing alcohol is also sold in concentrations higher than 70%, and as ethyl alcohol. Label it flammable and poisonous. Be sure your students clearly understand that they should never drink or taste anything unless specifically directed to do so.
☐ 100% corn oil.
☐ Baby oil (mineral oil).
☐ Dispensing containers. Use large baby food jars (6 oz. size) with eye droppers and lids.
☐ Soap and water to clean the 10 mL graduate of oil residue.
☐ Paper towels and masking tape. The

Saturated Salt Water   *Rubbing Alcohol Flammable - Poison*   *Corn Oil*   *Baby Oil*

rolled towels that students construct should be saved. They will dry overnight and can be reused many times.

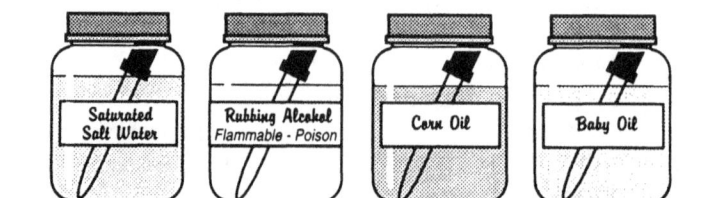

(TO) correlate the floating and sinking properties of 4 different liquids with their density measurements.

LIQUID PAIRS ◯ **Floating and Sinking ( )**

1. There are 6 different ways to pair these 4 liquids: water, rubbing alcohol, baby oil, corn oil. List these pairs on 6 lines of notebook paper.

1. water/rubbing alcohol: one layer
2. _____
3. _____
4. _____
5. _____
6. _____

2. Mix a few drops of each pair, one pair at a time, on the bottom of an inverted jar.

   a. If one layer forms, write this next to the liquid pair in your list.

   b. If two layers form, write this next to the liquid pair. Further, record which liquid floats on top and which sinks underneath.

   c. Wipe the glass dry with a paper towel after each test. Use just 1 towel to conserve trees.

3. How does density determine the floating and sinking behavior of each liquid?

© 1995 by TOPS Learning Systems

3

Answers / Notes

1-2.   water/ rubbing alcohol: one layer
     water/ baby oil: two layers; water sinks and baby oil floats
     water/ corn oil: two layers; water sinks and corn oil floats
     rubbing alcohol/ baby oil: two layers; rubbing alcohol sinks and baby oil floats
     rubbing alcohol/ corn oil: two layers; rubbing alcohol floats and corn oil sinks
     baby oil/ corn oil: one layer

3. When layers form, the liquid with the lower density always floats above the liquid with the higher density.

Materials

☐ A glass of water and an eye dropper.
☐ These 3 liquids, each in its dispensing container with an eye dropper:
      rubbing alcohol,
      baby oil (mineral oil),
      corn oil.
   Saturated salt water is not used here, but has wide application in later activities.
☐ A baby food jar or other small jar with a concave bottom. A laboratory watch glass or crucible is also suitable.

(TO) predict what floats and what sinks when water, rubbing alcohol, baby oil and corn oil are combined in various ways. To test these predictions by experiment.

PREDICTING LAYERS ○ Floating and Sinking ( )

1. Draw a diagram to *predict* what a test tube will look like if you gently add two droppers full of each liquid in this order:

    a. Explain your reasoning.
    b. Test your prediction: tilt the tube so each addition runs down its side.

1st — corn oil
2nd — water
3rd — alcohol
4th — baby oil

2. Draw a diagram to *predict* how the test tube will change if you now gently mix the liquids.

    a. Explain your reasoning.
    b. Test your prediction: cover the mouth with your thumb and slowly invert the tube *just once.*

LET EACH LIQUID FLOW GENTLY DOWN THE SIDE.

3. Vigorously shake the test tube to mix the liquids.
    a. What happens?
    b. Do the liquids stay mixed? Explain.

© 1995 by TOPS Learning Systems

Answers / Notes

1.

baby oil
rubbing alcohol
corn oil
water

1a. Corn oil is introduced first; next water, which sinks below it; next rubbing alcohol, which floats above the corn oil; finally baby oil, which floats above the rubbing alcohol. Liquids that would otherwise mix are separated by layers that don't mix.

1b. Students should evaluate their predictions. *The liquids must be poured down the side of the test tube as directed. Otherwise the rubbing alcohol might plunge through the corn oil barrier to mix with water. Or baby oil might break through the rubbing alcohol layer to mix with corn oil. Students who predict a different ordering should reexamine their density calculations.*

2.

baby oil and corn oil
water and alcohol

2a. If the oils come in contact, they should mix together; likewise the water and rubbing alcohol. As a result, only two layers will form in the test tube. Since water and alcohol have a higher average density than corn oil and baby oil, the water-alcohol layer should sink below the oil layer.

2b. Students should evaluate their predictions. *The liquids must be mixed gently as directed. Too much agitation will break the solution into a suspension of many bubbles.*

3. When the liquids are vigorously shaken together, a single opaque layer results. It is composed of many tiny water-alcohol bubbles dispersed between many tiny oil bubbles. Over time, bubbles of the same substance merge to again form two distinct layers.

Materials

☐ A test tube. Use a small jar or lump of clay to hold it upright. Or substitute the 10 mL graduate.
☐ A glass of water with an eye dropper.
☐ Three liquids in their dispensing containers: rubbing alcohol, baby oil and corn oil.

(TO) calculate the density of a candle by displacing water in a 100 mL graduated cylinder.

DUNK THE CANDLE ⭘ Floating and Sinking ( )

1. Get a candle with both ends cut square. Find its mass on a centered balance.

2. Bend the head of a pin with pliers to form a hook. Push it into the end cut most evenly. Slide the candle (hook up) into a 100 mL graduate about 1/3 full of water.

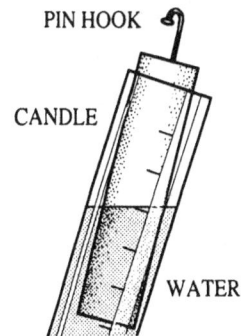

PIN HOOK

CANDLE

WATER

   a. Is the density of candle wax greater than 1.00 g/mL, or less? How do you know?

   b. Punch a hole in the end of a straw. Use the hole in this straw to "fish" the candle back out of the water by its hook.

3. Adjust the water level to 40.0 mL, then lower the candle back into the cylinder. Push on the hook with the straw to dunk *all* of the candle under water.

   a. How much water does the candle *displace* (push out of the way)? Find the difference between initial and final water levels.

   b. What is the volume of the candle? Explain your reasoning.

   c. Calculate the density of candle wax in g/mL.

   (Save your candle with hook and "fishing" straw to use again.)

© 1995 by TOPS Learning Systems   5

Answers / Notes

Numbers in our sample answers should correspond (within the limits of measuring error) to answers supplied by your students. Until now. Your particular candles will likely have masses much different than our candle of 28.1 g; volumes much different than 30.5 mL. Because different candle brands may have different wax compositions, your densities may also vary from our value of 0.92 g/mL by more than measuring uncertainty. Always use caution when applying our model answers (here and elsewhere) to the results your students reach; densities will usually match or be close, but masses, volumes and weights will vary.

1. *If you are using a TOPS balance, bend the paper clip on the object cup outward to make a supporting arm. This will prevent the candle from toppling out of the cup.*

   Mass candle = 28.1 g.

2a. The density of the candle is less than 1.00 g/mL, because it floats in water of this density.

3a.   Final water level (candle submerged): 70.5 mL
          Initial water level (no candle): 40.0 mL
             Volume of displaced water: 30.5 mL

3b. Volume of candle = 30.5 mL. When completely submerged, the candle occupies the same volume of space as the water it pushed out of the way.

3c. Density of candle = 28.1 g / 30.5 mL = 0.92 g/mL.

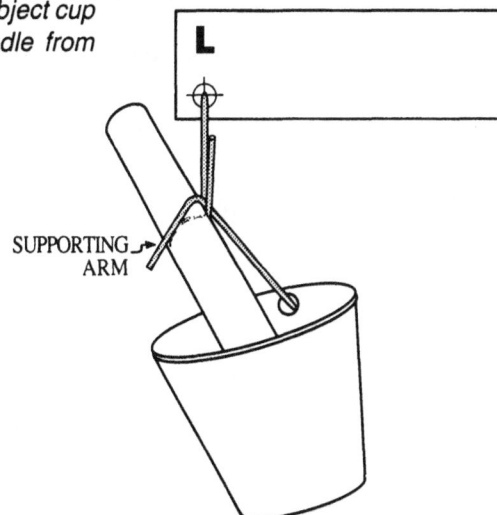

L

SUPPORTING ARM

Materials

☐ A utility (household/emergency) candle, 1.9 to 2.3 cm in diameter (3/4 to 7/8 inches), and 10 to 15 cm in length (4 to 6 inches), with ends cut squarely. It should be as large as possible and still fit inside a 100 mL graduate.
☐ A gram balance.

☐ A straight pin and pliers.
☐ A 100 mL graduated cylinder.
☐ A straw and paper puncher.
☐ A hand calculator.

(TO) compute the specific gravity of a candle using two different methods.

SPECIFIC GRAVITY ○ Floating and Sinking ( )

1. Gently lower your candle by its hook into a 100 mL cylinder full of water so it floats in the hump at the top. Catch overflow in a saucer.

2. Stick a pin into the floating candle precisely where the surface of the water touches its side. Lift the candle out.

3. Drill the pinhole a little larger, and smear a bit of clay in the hole. This marks its floating waterline.

    a. Cut, roll and tape notebook paper around that part of the candle that floats *under* water.

    b. Label it like this:

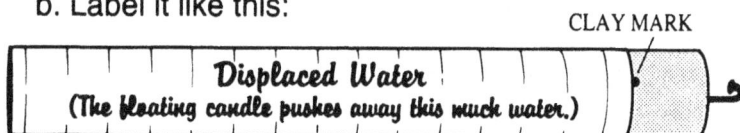

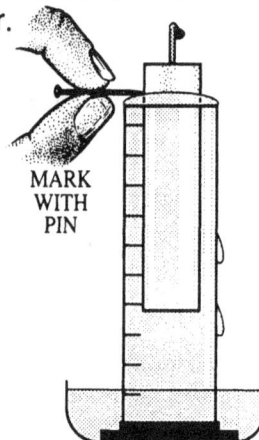

MARK WITH PIN

CLAY MARK

Displaced Water
(The floating candle pushes away this much water.)

4. <u>S</u>pecific <u>G</u>ravity (S.G.) compares the density of any substance to the density of water. Calculate the specific gravity of the candle in two different ways:

    a. $\text{S.G.} = \dfrac{\text{Density of candle}}{\text{Density of water}}$
    b. $\text{S.G.} = \dfrac{\text{Length of displaced water}}{\text{Length of candle}}$

6

Answers / Notes

1. *In this activity students measure water displacement by how far the candle sinks, not by the volume of water it displaces. Any deep, narrow container besides the 100 mL graduate could be substituted because length, not volume, is the measured variable.*

    The candle floats in the hump of water formed by surface tension; it must not be dunked. If it is thinner than the cylinder, it will lean a little, creating perhaps a 1 mm difference between the shorter and longer sides that are above water. This difference can be averaged by marking midway between these extremes, or ignored.

3a. *Parallel lines on the notebook paper help students estimate where to cut the paper.*

4a. Specific Gravity = $\dfrac{0.92 \text{ g/mL}}{1.00 \text{ g/mL}} = .92$

4b. Specific Gravity = $\dfrac{9.75 \text{ cm}}{10.80 \text{ cm}} = .90$ *If the candles' ends are uneven, students can minimize this error by measuring through the clay mark to find both the length of the whole candle and its submerged length.*

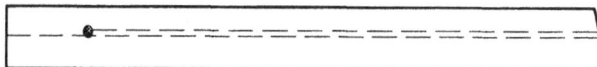

Materials

- The utility candle with hook, and "fishing" straw from the previous activity.
- A 100 mL graduate and an overflow margarine tub or equivalent.
- A pin.
- A small lump of oil-based modeling clay.
- Notebook paper.
- Scissors.
- Tape.
- A calculator.

(TO) observe that a floating candle displaces a mass of water equal to its own mass.

FLOAT THE CANDLE ⭕   Floating and Sinking ( )

1. Slide the wrapper off your candle. Check that your clay dot still marks the candle's float line in a 100 mL graduate brimful of water.

2. Now fill the graduate with 40.0 mL of water. Gently float the candle, without splashing.

   a. Read the volume where the clay mark floats. (This is the true float line, even if water creeps higher in the narrow space between candle and cylinder wall.)

   b. Find the difference between initial and final volumes. This displaced water equals the volume of your paper cylinder.

3. You know that water has a density of 1.00 g/mL. This means that each mL of displaced water has a mass of 1 gram.

   a. What is the total mass of water represented by your paper cylinder?

   b. Recall the total mass of your candle from a previous activity. Compare this value to the mass of its displaced water.

   c. What seems to be the relationship between the mass of a floating candle and the mass of water it displaces?

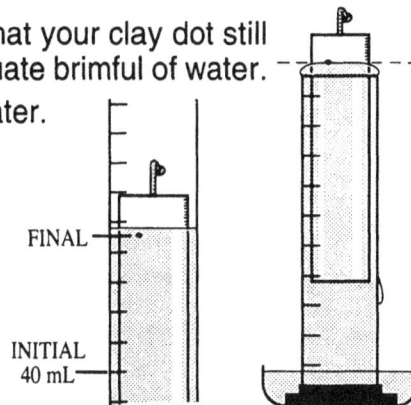

FINAL

INITIAL 40 mL

     7

Answers / Notes

1. *If the dot is not accurately placed, students should scrape it away and apply a more accurate mark.*

2a. *"Water creep" is more formally called capillary action, a tendency of water to rise into and fill narrow or confined space. Our candle floated at a level where its clay dot matched 67.5 mL.*

2b.   Final volume = 67.5 mL
      Initial volume = 40.0 mL
      Volume displaced water = 27.5 mL

3a. Mass displaced water = 27.5 mL x 1 g/mL = 27.5 g

3b. In activity 5, the mass of our candle was determined to be 28.1 g. Within the limits of measuring uncertainty, this equals the mass of water it displaces.

3c. The floating candle apparently displaces a mass of water equal to its own mass. *(While this is true only for experiments performed on Earth, it is everywhere true that floating bodies displace their own weight in water.)*

Discussion

The *mass* of a candle (the amount of matter it contains) is everywhere the same. The *weight* of a candle (its gravitational attraction) changes from place to place. It is the downward force of the candle's *weight*, not its *mass*, that displaces (pushes away) water.

Mass is measured in grams on a balance beam. Weight is measured in ounces (or newtons) on a spring scale. Among earthbound scientists in a constant gravitational field, the numbers on both instruments rise and fall in direct proportion. They are experimentally equivalent, though conceptually distinct. Thus far, for accuracy and convenience, we have measured only mass on gram balances. In later activities (15-18) we will use spring balances to develop Archimedes principle in terms of weight.

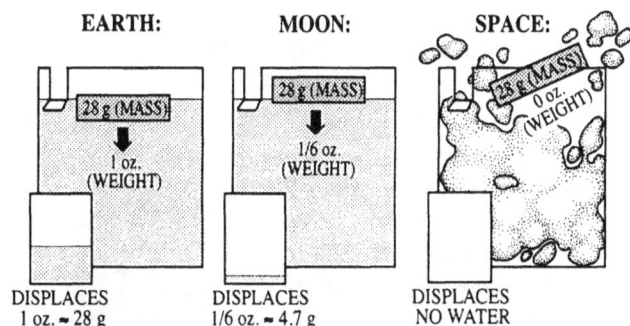

EARTH: MOON: SPACE:

28 g (MASS)   28 g (MASS)   28 g (MASS)
0 oz. (WEIGHT)

1 oz. (WEIGHT)   1/6 oz. (WEIGHT)

DISPLACES 1 oz. ≈ 28 g   DISPLACES 1/6 oz. ≈ 4.7 g   DISPLACES NO WATER

Materials

☐ The utility candle with pin hook, paper wrapper, and float line marked with clay.

☐ The "fishing" straw with a hole punched in the end.

☐ A 100 mL graduate.

☐ An overflow tub or equivalent.

(TO) apply concepts developed in the study of a floating candle (density, specific gravity and water displacement) to a floating wood dowel.

HOW NOW, BROWN DOWEL? ○   Floating and Sinking (   )

Get a large wood dowel with both ends cut square.

1. Find its density.

2. Find its specific gravity using 2 different ratios.

3. Show that it displaces a mass of water equal to its own mass.

Review these 3 task cards!

DUNK THE CANDLE   5
SPECIFIC GRAVITY   6
FLOAT THE CANDLE   7

© 1995 by TOPS Learning Systems

8

Answers / Notes

1. *This is a review of activity 5:*

Attach a pin hook to the dowel. (Use pliers to push it in.) Push on this hook to dunk the wood *completely* under water:

final water level = 83.3 mL
initial water level = 40.0 mL
volume of water displaced = 43.3 mL
volume of dowel = 43.3 mL

Blot water off the dowel and find its mass on a gram balance. *(Unlike candle wax, wood does absorb some water. Its "damp" mass is therefore the best estimate of density while floating)*:

mass of wood dowel = 33.5 g
density of dowel = 33.5 g / 43.3 mL = 0.77 g/mL

2. *This is a review of activity 6:*

Float the dowel in a 100 mL graduate that is filled to the brim with water. Mark the float level with a pencil. Don't count the small lip of water that clings to the wet wood because of surface tension. Calculate its <u>S</u>pecific <u>G</u>ravity as a ratio of density and a ratio of underwater length to full length:

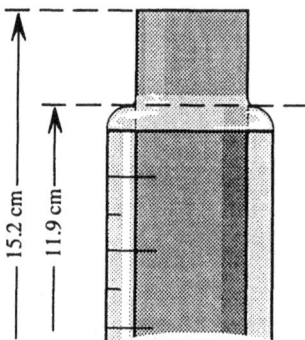

15.2 cm   11.9 cm

density ratio: S.G. = 0.77 g/mL / 1.00 g/mL = 0.77 g/mL
length ratio: S.G. = 11.9 cm / 15.2 cm = 0.78 g/mL

3. *This is a review of activity 7:*

Measure how much water is displaced by the *floating* dowel in a graduated cylinder:

final float level at pencil mark = 73.8 mL
initial water level = 40.0 mL
volume of displaced water = 33.8 mL
mass of displaced water = 33.8 g
mass of wood dowel = 33.5 g

Within the limits of experimental error, these last two values are equal. The dowel displaces a mass of water equal to its own mass.

Materials

☐ A wood dowel with squarely cut ends, similar in size to the utility candle used in the last three activities. Again, it should be as large as possible and still fit inside a 100 mL graduate: 1.9 to 2.3 cm in diameter (3/4 to 7/8 inches), and 10 to 15 cm in length (4 to 6 inches).
☐ Two straight pins.
☐ Pliers.
☐ A gram balance.
☐ A 100 mL graduate and overflow tub.
☐ A calculator.

(TO) experimentally determine the density of oil-based clay. To review all density data collected thus far and confirm its accuracy.

THE NATURAL ORDER ○ Floating and Sinking (   )

1. Roll precisely 50.0 g of clay into a cylinder shape so it fits inside a 100 mL graduate. Stick a bent pin hook in one end that you can catch in the hole of your "fishing" straw as before.

   a. Use water displacement to find the volume of this clay.

   b. Calculate its density.

2. You know the densities of 5 liquids and 3 solids. List these substances in order, from least dense to most dense, along with your experimental data.

Least Dense
_____
_____
_____
_____
_____
Most Dense

3. Build up layers of liquids and solids in a test tube or 10 mL graduate. Did everything behave as you expected? Explain.

BIRTHDAY CANDLE
TOOTHPICK
● CLAY

© 1995 by TOPS Learning Systems   9

Answers / Notes

1. *Students should add or subtract bits of clay to a gram balance until they assemble a 50.0g mass, then roll it into a cylinder shape. If the clay is difficult to manipulate, warm it under hot tap water.*

1a. final water level with clay = 70.6 mL
   initial water level = 40.0 mL
   volume clay cylinder = 30.6 mL

1b. density clay = 50.0 g / 30.6 mL = 1.63 g/mL

2-3. *The last figure in each density measurements is uncertain. Expect the results of your students to differ from these by plus or minus 0.02 g/mL. Because of this uncertainty, our density data did not predict, with certainty, whether candle wax floats or sinks in corn oil.*

3. The liquids and solids assemble into layers as shown, less dense substances floating above more dense substances. There were, as well, some unexpected results:

   As fresh water runs down the side of the test tube into salt water, you can actually observe *(because of light interference)* that it briefly floats on top before mixing. This confirms, as our data suggests, that fresh water is less dense than salt water.

   Another surprise is that the small piece of birthday candle *just* floats rather than sinks in corn oil. *(Surface tension may cause the candle to float unusually high. Push it under the corn oil with a straw to observe how slowly it rises and how low it really floats. Different brands of candles and corn oil may produce opposite results.)* Corn oil is slightly denser than candle wax, though measuring uncertainty in our data didn't allow us to detect this.

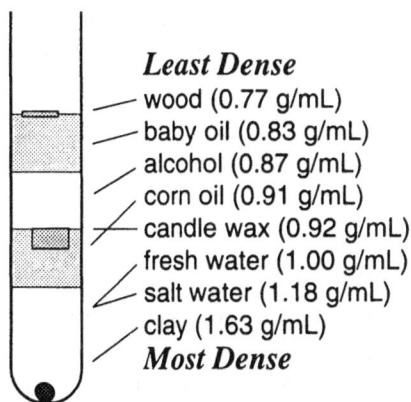

Least Dense
— wood (0.77 g/mL)
— baby oil (0.83 g/mL)
— alcohol (0.87 g/mL)
— corn oil (0.91 g/mL)
— candle wax (0.92 g/mL)
— fresh water (1.00 g/mL)
— salt water (1.18 g/mL)
— clay (1.63 g/mL)
Most Dense

Materials

- ☐ Modeling clay.
- ☐ A gram balance.
- ☐ A 100 mL graduate.
- ☐ A straight pin and pliers.
- ☐ The "fishing" straw with hole punched in the end.
- ☐ A calculator.
- ☐ Five liquids: saturated salt water, fresh water, corn oil, rubbing alcohol, baby oil.
- ☐ Three solids: a small ball of clay, pieces of toothpick and birthday candle.
- ☐ A test tube with jar "holder," or a 10 mL graduate.

(TO) determine the volume of a golf ball using an overflow cup. To observe that a milliliter occupies the same volume as a cubic centimeter.

OVERFLOW CUP ○ Floating and Sinking ( )

GOLF BALL

1. Make 2 parallel cuts, 1 cm wide, about 1/4 of the way down the side of a paper cup. Fold out the flap to make a spout on an *overflow cup.*

2. Set this cup on an inverted can beside a plastic tub. Fill it with water until it overflows into the tub.

3. Replace the tub with a large graduated cylinder. Gently drop a golf ball into the cup and measure all the displaced water that overflows into the graduate.

4. What is the volume of the golf ball in mL? Explain how you know.

5. Set the golf ball between 2 batteries. Measure its diameter (the distance between batteries) with a centimeter ruler. What is its radius?

6. The volume of a ball in <u>cm³</u> is given by $V = 4/3\pi r^3$. Use a calculator to find this volume.

BALL'S DIAMETER

7. Compare your volumes in steps 4 and 6. What can you say about the relationship between milliliters (liquid measure) and cubic centimeters (dry measure).

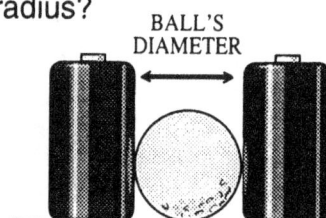

© 1995 by TOPS Learning Systems

10

Answers / Notes

4. The golf ball sinks, displacing a volume of water equal to its own volume. When this overflow is captured in a graduated cylinder, it measures about 40.3 mL. *(Student answers may vary, because the surface tension of water doesn't always stop its flow from the cup at precisely the same level.)*

5. Distance between batteries = diameter golf ball = 4.25 cm; radius = 4.25 cm/2 = 2.13 cm

6. $V = (4/3)(3.14)(2.13 \text{ cm})^3 = 40.5 \text{ cm}^3$

7. Within the limits of measuring error, both volumes are equal: 40.3 mL ≈ 40.5 cm³. Thus milliliters and cubic centimeters occupy the same space: 1 mL ≈ 1 cm³.

   Expect student variations to be wider than this. A 1 mm range of diameter uncertainty (4.20 — 4.30 cm) produces almost a 3 cm³ range of volume uncertainty (38.8 — 41.6 cm³).

Materials
- ☐ A paper drinking cup, 6 oz. or larger. Don't substitute styrofoam; it breaks when the pour spout is folded.
- ☐ Scissors.
- ☐ An empty can to support the overflow cup.
- ☐ A jar or glass of water to fill the overflow cup.
- ☐ A tub to catch the initial overflow.
- ☐ A golf ball.
- ☐ A 100 mL graduated cylinder.
- ☐ Two size-D batteries, dead or alive.
- ☐ A centimeter ruler.
- ☐ A calculator.

(TO) calculate mass knowing density and volume. To calculate volume knowing density and mass.

DENSITY MATH ○   Floating and Sinking (   )

1. Roll a lump of clay that has about the same volume as a size-D battery.

   a. Pull off about 1/3 of this clay and shape it into a cube. Shape the remaining 2/3 like a pyramid.

   b. You have previously determined the density of this clay. What is it?

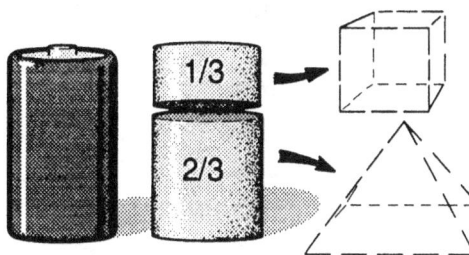

2. Find the volume of the pyramid with your overflow cup.

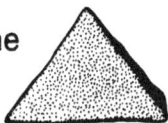

   a. Multiply the pyramid's volume by its density to find its mass.

      volume x density = mass
      mL x g/mL = g

   b. Check this result on your balance.

3. Find the mass of the cube on your balance.

   a. Divide the cube's mass by density to find its volume.

      mass/density = volume
      g x mL/g = mL

   b. Check this result with an overflow cup.

© 1995 by TOPS Learning Systems   11

Answers / Notes

1b. Density of clay = 1.63 g/mL.

2. Volume of water displaced by pyramid in overflow cup = 26.5 mL
2a. 26.5 mL x 1.63 g/mL = 43.2 g
2b. Actual mass on balance = 43.6 g (good agreement)

3. Mass of clay on balance = 26.1 g
3a. 26.1 g x 1 mL/1.63 g = 16.0 mL
3b. Actual volume using overflow cup = 16.0 mL (excellent agreement)

Materials
☐ Modeling clay.
☐ A size-D battery, dead or alive.
☐ Density results from activity 9.
☐ An overflow cup with support can and water tub.
☐ A 100 mL graduate.
☐ A calculator.
☐ A gram balance.

(TO) mold clay into a floating shape that displaces its mass in water. To reshape the same piece of clay into a sinking shape that displaces its volume in water.

CLAY BOAT   O   Floating and Sinking ( )

1. Mold a 50.0 g mass of clay into a flat "pancake."

2. Wrap a size-D battery, *from the bottom up*, in a square of plastic wrap.

   a. Mold your clay around the battery, *from the bottom up*. Extend the sides nearly to the top.

   b. Remove the battery and pull out the plastic.

   c. Hold your clay "boat" up to bright light to detect holes. Fix them so it floats in water with no leaks.

   d. Check that your boat still has a mass of 50.0 g.

3. What volume of water will your clay boat displace in an overflow cup? *First* make a reasoned prediction, then test.

4. What volume of water will this clay displace if you squash it together and roll it into a sphere? *First* make a reasoned prediction, then test.

© 1995 by TOPS Learning Systems   12

Answers / Notes

1. *If this much clay is difficult to shape, warm it under hot tap water.*

3. Prediction: The clay boat floats in the overflow cup. Thus it displaces a 50.0 g mass of water equal to its own 50.0 g mass. Since each gram of water has a volume of 1 mL, 50.0 mL of water will flow out of the cup.

   Test: Our clay boat displaced 49.1 ml of water into a graduated cylinder, reasonably close to the predicted value.

4. Prediction: If you squash the clay into a sphere it sinks in the overflow cup. Thus it displaces a volume of water equal to its own volume. Knowing the mass and density of the clay sphere, we can calculate how much water will overflow when it is placed in a brimful overflow cup.

$$\text{volume clay sphere} = 50.0 \text{ g} \times 1\text{mL} / 1.63 \text{ g} = 30.7 \text{ mL.}$$

   Test: Our clay sphere displaced 30.0 mL of water into a graduated cylinder, close to the predicted value.

Materials

☐ Modeling clay. Use the 50.0 g clay cylinder from activity 9, if available, and skip this initial mass determination.
☐ A gram balance.
☐ A size-D battery.
☐ Plastic wrap.
☐ An overflow cup with support can and water tub.
☐ A 100 mL graduated cylinder.
☐ A calculator.

(TO) cover a cork with clay until its density is slightly greater than fresh water. To float this body on a layer of salt water submerged under fresh water.

NEITHER HERE NOR THERE ○   Floating and Sinking (   )

1. Cover the cork supplied by your teacher with a thin "pancake" of clay until you can't see the cork.

2. Put this clay/cork mass in a glass of water at room temperature. If it floats, shake off the water and add more clay. If it sinks, scrape clay away.

CLAY-COVERED CORK

    a. Make fine adjustments until it *both* remains at the water's surface *and* sinks when you push it below the surface. Smooth the clay evenly around the cork when you find the right amount.

    b. What is the average density of your clay/cork? (No calculations necessary!)

3. Push your clay/cork to the bottom of the glass. It should remain there, not quite rising.

    a. Trickle eye droppers full of saturated salt water down the side of the glass, until the clay/cork neither sinks nor floats.

    b. Explain what's happening.

4. Save your clay/cork in a small plastic bag labeled with your name.

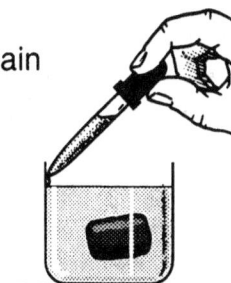

© 1995 by TOPS Learning Systems   13

Answers / Notes

2a. *This clay/cork "remains" at the surface of the water. It doesn't "float" in the scientific sense because it no longer displaces a mass of water equal to its own mass. Rather, the cohesive force of water, its tendency to cling to itself, holds the clay/cork up: the surface of the water is broken by water-repelling, oil-based clay and cohesion prevents the water from covering this "hole" back over.*

2b. The clay body has an average density that is slightly greater than 1.00 g/mL, the density of water.

3a. *This trick can also be accomplished with an egg.*

3b. The saturated salt water initially sinks to the bottom of the glass. This bottom layer of salt water is more dense than both the clay/cork and fresh water above it. The clay/cork rises until it displaces a mass of fluid exactly equal to its own mass. There it floats. *(Over several hours, the salt solution disperses throughout the entire glass of water, from bottom to top, until its density everywhere exceeds the density of the clay/cork. This raises it to float at the surface.)*

Materials

☐ A natural cork (not rubber) with a mass of about 2 grams. Look for a #9 size cork, or slice off a 2 gram piece of wine cork (about 2.4 cm long). Clay surrounding a cork of this size will perfectly match the capacity of spring scales your students will improvise in activity 15.

☐ Modeling clay.

☐ A glass or jar of room-temperature water. This should be deep enough to completely submerge the clay/cork, shallow enough to reach in and retrieve it with your fingers.

☐ Saturated salt water with an eye dropper dispenser.

☐ A plastic sandwich bag.

☐ Masking tape.

(TO) create 3 bodies of equal mass but unequal volume. To understand how volume affects the floating and sinking characteristics of a body.

THREE CLAY BODIES ◯ Floating and Sinking (   )

1. Get your clay/cork from its plastic bag. Find its mass on a gram balance.

   a. Get another cork, about twice as large as the first. Cover it with clay so it has precisely the same mass as your original clay/cork.

   b. Roll a ball of pure clay that has precisely the same mass as your large and small clay/corks.

ORIGINAL CLAY/CORK   LARGER CLAY/CORK   PURE CLAY

EQUAL MASS.

2. Put each clay body in water. Lightly press an F, N, or S into each one with your pencil to indicate whether it Floats, is Neutral, or Sinks in water.

3. Which of these 3 bodies displaces its own *mass* in water? Use your overflow cup and a 100 mL graduate to evaluate.

4. Which of these 3 bodies displaces its own *volume* in water? Explain your reasoning.

5. F, N, and S all have the same mass. Why, then, do they behave differently in water?

6. Store all 3 clay bodies in your labeled plastic bag.

© 1995 by TOPS Learning Systems   14

Answers / Notes

1. Our clay/cork had a mass of 17.1 g.

1a. *Students should set this larger cork on a centered balance, and add just enough clay to equal the mass of their smaller clay/cork. Then they should mold this clay into a thin "pancake," and cover the larger cork with a smooth, thin, even coat.*

3. Each clay body displaced these volumes of water in an overflow cup, and therefore these masses:

   F = 17.0 mL displaced water = 17.0 g
   N = 16.9 mL displaced water = 16.9 g ── *(If cohesion holds N above the surface in the overflow cup,*
   S = 14.3 mL displaced water = 14.3 g   *quickly push it under the water.)*

   In step 1, each clay body was formed with just enough clay to have a mass of 17.1 g. This value is in close agreement with F and N. Both of these bodies displace their own mass in water. S, by contrast, displaces less than its own mass in water.

4. N and S each displace their own volumes in water because they are completely under water. F, by contrast, displaces less than its own volume because it is only partially submerged.

5. These bodies of equal mass (M) behave differently in water because they have different volumes. *The Floating body's volume is large enough to displace M with above-water volume to spare. The Neutral body's smaller volume displaces M and achieves zero buoyancy, but with no volume left over to remain above water. The Sinking body's volume is smaller yet, displacing less than M.*

Materials

☐ The clay/cork in its labeled plastic bag from the previous activity.
☐ A gram balance.
☐ A larger natural cork, about twice as big as the former, with a mass somewhere near 4 grams. If you used a half wine cork in the last activity, use an undivided wine cork here.

☐ Modeling clay.
☐ A glass or jar of water.
☐ An overflow cup with support can and water tub.

(TO) construct a sensitive spring scale that measures force in "weight units." To provide an experimental basis for the study of buoyancy and Archimedes' principle.

SPRING SCALE   ○   Floating and Sinking ( )

1. Cut out a <u>Weight Scale</u>. Tape 4 short lengths of straw to it where shown.

2. Loop a rubber band near the side of a big cereal box. Fold your Weight Scale around the middle of that side, and tuck the straws under the rubber band.

3. Cut 1.5 meters of steel wire. Form a *very* tight knot 11 cm from one end. Coil it closely and evenly (without overlap) around your pencil, starting 5 cm beyond the knot.

4. Slide this spring off your pencil. Straighten the last coil. Tape it to the top of your box so the spring hangs over the scale.

5. Pull out the "arm" of a paper clip a little, to form a hook. Wire it to the bottom of your spring.

6. Tape a 15 cm thread "bucket handle" to a film canister. Hang it on the hook, over your table edge.

7. Stretch out the coils of your spring so they hang <u>halfway</u> down the box *when relaxed*. (Pull them a little past halfway, then release.)

8. Slide the scale down and out until the "0" line touches the knot. This centers your spring scale.

© 1995 by TOPS Learning Systems   15

Answers / Notes

7. If the spring gets stretched too far, reinsert the pencil through the coils and recompress them. You can do this without untaping the spring.

8. Build a spring scale in advance. Test it for these properties:

• SENSITIVITY: Each clay body (from activity 14) should weigh somewhere between 20 and 30 weight units (wu). If each body weighs less than 20 wu, substitute thinner wire, or wrap a longer piece of wire around the pencil to make additional coils. If each clay body weighs more than 30 wu, cut off about 1/6 of the coils.

• STRENGTH: Stretch the spring so the knot (centered at zero) moves through the full weight range, from 0 wu to 60 wu. Operating within this range, the spring should not stretch beyond its elastic limit: the knot should return to its centered position at the zero mark. If it settles lower, stretch out the spring a centimeter or two more. Slide the scale lower to recenter the knot at zero, and test again.

• CONSISTENCY: Different spring scales may register different weights, but all three clay bodies should consistently weigh the same on any particular scale. (In our particular case, each clay body weighs 25.0 wu.)

As long as students in each lab group limit all weighings to the same scale, calibration is unnecessary. Accuracy to the nearest 0.1 wu is possible, but demanding. We have rounded off to the nearest 0.5 wu in experiments that follow. Accuracy to the nearest 1 wu is also satisfactory.

Materials

☐ The Weight Scale pattern. Photocopy this supplementary cutout (4 to a page) at the back of this book.

☐ Scissors.

☐ A plastic drinking straw and masking tape.

☐ A cereal box at least 30 cm (1 foot) tall or taller. We used a 20 ounce Cheerios box.

☐ A rubber band, long enough to fit around the height of your cereal box.

☐ A standard wooden pencil (hexagonal, about 1/4 inch in diameter).

☐ Dark annealed or galvanized steel wire with a 28 gauge thickness. Thicker 24 gauge wire will not work unless you coil very long 5 meter pieces. Thinner 30 gauge wire is suitable if you wind it in tighter coils, around a 3/16 inch diameter straw.

☐ A meter stick.

☐ Wire cutters.

☐ A paper clip.

☐ Thread and tape.

☐ A film canister for 35 mm film.

(TO) study relationships between weight, buoyancy and water displacement for sinking, neutral and floating bodies.

FRESH WATER WEIGH-INS ○ Floating and Sinking ( )

1. Stick a pin through the sticky side of a masking tape square. Tape it on the bottom of the canister on your spring scale, point down.

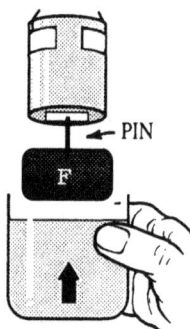

← PIN

2. Center the knot at "0" weight units by adjusting the movable scale
3. Stick clay body F on the pin.
   a. Weigh it in air; in a glass of water.
   b. Copy this table. Write your results in the top row of the first 2 columns.
4. Recenter your balance. Record weights for N and S in air and in water.

---

5. An object immersed in water is pushed up by a force called *Buoyancy*. Calculate the buoyancy of water on each body and fill in the third column.

$$B = (Wt)_{Air} - (Wt)_{Water}$$

FRESH WATER TABLE (Weight Units)

	Weight in AIR	Weight in WATER	BUOYANCY of Water	Weight Water DISPLACED
Body F				
Body N				
Body S				

6. Collect the water displaced by each body in your overflow cup, and weigh it. Fill in the fourth column.
7. Within the limits of experimental error, what generalizations can you make?

16

Answers / Notes

3-6. *Each body can be weighed inside the canister equally well, without sticking it on the pin. But weighing it in water then becomes a separate step that still requires use of the pin.*

Stress the importance of always checking to be sure the spring scale is still centered before each new weighing. The sliding scale is easily knocked out of alignment, and the spring can be easily stretched beyond its elastic limit.

FRESH WATER TABLE (Weight Units)

	Weight in AIR	Weight in WATER	BUOYANCY of Water	Wt Water DISPLACED
Body F	25.0 wu	0 wu	25.0 wu	25.0 wu
Body N	25.0 wu	0 wu	25.0 wu	25.0 wu
Body S	25.0 wu	10.5 wu	14.5 wu	14.5 wu

7. Floating and neutral bodies displace their own weights in water. A sinking body displaces less than its own weight in water.

Any immersed body (floating, neutral or sinking), is buoyed up by a force equal to the weight of water it displaces. *(This principle was first articulated by Archimedes, a Greek philosopher, in the 3rd century BC.)*

Materials

☐ The spring scale previously constructed.
☐ A pin.
☐ Masking tape.
☐ Three clay bodies in a plastic bag.
☐ An overflow cup with support can and water tub.
☐ A glass of fresh water.

(TO) study relationships between weight, buoyancy and liquid displacement for sinking and floating objects in liquids of different densities.

SALT WATER WEIGH-INS ○

1. Copy and complete this table for *salt water* as you did for fresh water.

SALT WATER TABLE (Weight Units)

	Weight in AIR	Weight in SALT WATER	BUOYANCY of Salt Water	Wt Salt Water DISPLACED
F				
S				

3. Compare *fresh* water data (from before) and this salt water data for the <u>S</u>inking body:

　　a. Are equal weights of both liquids displaced? Why?

　　b. Does S weigh the same in both liquids? Why?

Floating and Sinking ( )

2. Read Archimedes' principle. Does it hold for salt water? Explain.

> ### Archimedes' Principle
> An immersed object is buoyed up by a force equal to the weight of the fluid it displaces.

4. Compare *fresh* water and *salt* water data for the <u>F</u>loating body:

　　a. Does F weigh the same in both liquids? Why?

　　b. Would you expect the volumes of each displaced liquid to be equal? Predict, then test.

FRESH WATER　　SALT WATER

© 1995 by TOPS Learning Systems　　17

Answers / Notes

1. *N is not included in this table since it cannot be <u>N</u>eutral in both solutions.*

2. Yes. Archimedes' principle predicts that any object (floating or sinking) immersed in any liquid is buoyed up with a force that is equal to the weight of liquid it displaces. Within the limits of experimental error, this is true: buoyancy (in the third column) matches the weight of displaced salt water (in the fourth column).

SALT WATER TABLE (Weight Units)

	Weight in AIR	Weight in SALT WATER	BUOYANCY of Salt Water	Wt Salt Water DISPLACED
F	25.0 wu	0 wu	25.0 wu	24.0 wu
S	25.0 wu	7.5 wu	17.5 wu	17.5 wu

3a. No. The sinking body displaces more salt water (17.5 mu) than fresh water (14.5 wu). It displaces equal volumes of both liquids, but the higher density salt water still weighs more.

3b. No. The sinking body weighs 10.5 wu in fresh water, but only 7.5 wu in salt water. It weighs less in salt water because it displaces a greater weight of salt water, and hence is buoyed up with greater force. *(Anyone who goes swimming in very salty water — the Great Salt Lake or Dead Sea, for example — will immediately feel the extra buoyancy exerted by these superdense waters. You need not tread water to keep your head above the surface; you simply float.)*

4a. Yes. The full weight of F is buoyed up by both liquids. It weighs 0 mu in both fresh water and salt water.

4b. No. S must sink deeper into lower-density fresh water before it displaces its own weight.

　This, in fact, is the case. The floating body displaces 17.0 mL of fresh water from an overflow cup, but only 14.4 mL of salt water.

Materials

☐ The spring scale and clay bodies.

☐ The overflow cup with support can and water tub.

☐ A graduated cylinder.

☐ Saturated salt water. Each lab group will likely require more salt water than you may have dispensed in the baby food jar of activity 2. Prepare a fresh batch, if necessary. Or direct students to reuse salt water by pouring it from their overflow cups back into its holding jar.

☐ The fresh water data table from activity 16.

(TO) verify Archimedes' principle with a floating candle and a sinking rubber stopper.

ARCHIMEDES' PRINCIPLE ○ Floating and Sinking ( )

1. Fill the canister on your spring scale brimful of water. Put a tub underneath to catch the overflow.

a. If you float a piece of a candle in the canister, will the knot move lower? Use Archimedes' principle to support your answer.

b. Test your prediction.

KNOT

CANDLE

BRIMFUL

TOTAL WEIGHT

a.

DISPLACED WATER

b.

2. Dry the canister and recenter your spring scale to "0".

a. Drop a rubber stopper into an overflow cup brimful of water. Catch all water it displaces in your canister.

b. Hang the stopper plus displaced water on your scale. Buoy up the stopper in a glass of water, and read the total weight on your spring scale.

c. Does this weight equal the stopper's weight? Use Archimedes' principle to support your answer.

d. Test your prediction.

© 1995 by TOPS Learning Systems   18

Answers / Notes

1a. No. The knot on the spring scale will not move. Archimedes' principle predicts that the candle is buoyed up by a force equal to the weight of the displaced water that spills over the top. The candle adds weight to the scale in the same amount that the overflowing water subtracts weight.

1b. *Students should float the candle in the brimful canister and observe that the weight remains virtually unchanged. Ask them to rethink incorrect predictions.*

2b. weight of stopper buoyed by water + displaced water = 17.5 weight units.

2c. This total weight (in our case, 17.5 wu) will be equal to the weight of the stopper alone. Archimedes' principle predicts that the stopper is buoyed up by a force equal to the weight of the displaced water in the canister above it. The buoyancy of water lifts the stopper with a force that is equal to the weight of the displaced water pushing down. These forces cancel, leaving only the weight of the stopper registered on the scale.

2d. *Students should weigh the stopper alone and observe that the scale registers nearly the same weight (in our case, 17.0 wu). Ask them to rethink incorrect predictions.*

Extension

• Show that Archimedes' principle holds in step 1 for salt water.

• Show that Archimedes' principle holds in step 2 for a floating candle. (Suspend the candle beneath the canister from thread, not its pin. Otherwise the canister of displaced water will "stand" on the floating candle and be partially supported by it.)

Materials

☐ The spring scale.
☐ A tub to catch overflow from the canister and overflow cup.
☐ A chunk of utility candle cut small enough to fit under the thread handle and float in the film canister.
☐ The overflow cup with support can.
☐ A solid rubber stopper.
☐ A glass of water.

(TO) make a hydrometer. To explain how it measures density.

HYDROMETER (1)   ○   Floating and Sinking (   )

1. Close one end of a straw with a small plug of clay.

2. Melt a puddle of wax onto a plastic lid from a lighted candle (about a dozen drops). When the liquid starts to solidify, dip in the plugged end to seal with wax.

3. Set a 100 mL graduate, filled to the top with water, in an overflow saucer. Add BB's to your plugged straw so only 1/5 of it floats above water.

4. Close the straw with masking tape. Trim off excess tape.

5. Your straw should now float in a "hump" of water above the graduate's mouth. If not, add more water.

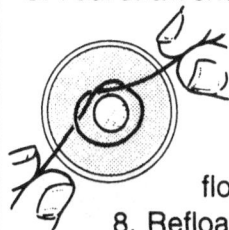

6. Lower a dry loop of thread over the straw until it floats on this hump. Pull it tightly around the straw to mark the float line.

7. Lift the straw from the water. Mark the float line with pencil, then remove the thread.

8. Refloat the straw in the hump of water to check your float line for accuracy. If necessary, erase and remark.

9. You have just made a hydrometer, an instrument that measures density. How do you think it works?

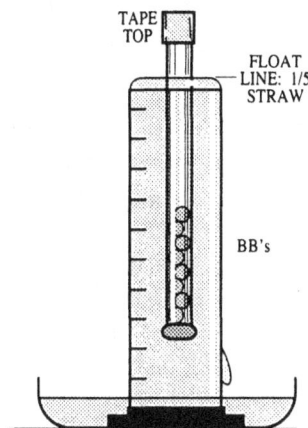

TAPE TOP

FLOAT LINE: 1/5 STRAW

BB's

19

Answers / Notes

2. *It is important to wait until the wax begins to solidify. Partial cooling prevents it from melting the clay plug. The clay and wax seal must be airtight. Otherwise water will seep into the straw and gradually sink it lower into the water. Check this by attempting to blow air through the straw into water. You should see no bubbles.*

4. *This tape top is not airtight, and must be kept dry. Its only purpose is to prevent the BB's from rolling out when the hydrometer is laid on a horizontal surface.*

6. *Form a loose loop with a "half knot" to drop over the straw. Dry thread floats on the surface of the water, accurately marking the water level when pulled tight. Wet thread may sink.*

9. The resting position of the hydrometer in liquid gives a relative measure of density. It sinks deeper into liquids that are less dense and floats higher in liquids that are more dense.

Materials

- ☐ A straight plastic drinking straw.
- ☐ Modeling clay.
- ☐ A candle and matches.
- ☐ A plastic lid.
- ☐ A 100 mL graduate.
- ☐ A tub to catch overflow from the graduate.
- ☐ BB shot.
- ☐ Masking tape.
- ☐ Scissors.
- ☐ Thread.

(TO) calibrate a hydrometer. To use these calibrations to estimate the density of an unsaturated salt water solution.

HYDROMETER (2)   ◯   Floating and Sinking (   )

1. You already marked the float line for water on your hydrometer at 1.00 g/mL.
   a. Recall the density of saturated salt water. Mark a new float line for this liquid by repeating steps 5-7 from the last activity.
   b. Refloat your hydrometer in a "hump" of salt water to check that this new float line is accurately placed. If not, erase and remark.

2. Cut a rectangle of paper with an edge equal to the distance between both density values on your hydrometer. Label it like this:

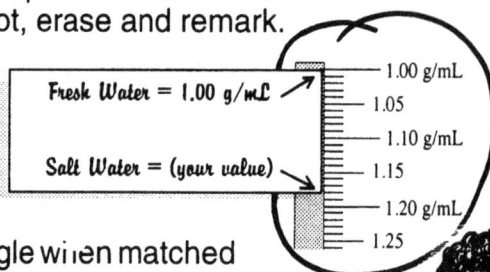

   Fresh Water = 1.00 g/mL
   Salt Water = (your value)

   1.00 g/mL
   1.05
   1.10 g/mL
   1.15
   1.20 g/mL
   1.25

3. Get a sheet of <u>Hydrometer Scales</u>. Find the scale that best fits this rectangle when matched against your pair of density values. Circle it.

4. Dissolve a tablespoon of salt in a glass of water.
   a. Measure the density of this solution with your hydrometer and scale. Tell how you did this.
   b. Find its density by another method and compare values. Which method do you think is most reliable, a or b? Explain.

© 1995 by TOPS Learning Systems   20

Answers / Notes

1a. Our salt water solution had a density of 1.18 g/mL.

2-3. *Although this paper rectangle simplifies the search for a scale of the correct width, it may be omitted by using the straw itself to search for a match.*

4a. Fill a 100 mL graduate brimful with the salt solution of unknown density. Mark its float level with a thread and pencil as before. Line up the highest and lowest marks on the hydrometer to match the fresh water and saturated salt water densities on the circled scale. Read the density in between that corresponds to the pencil mark for this new solution.

    The pencil mark on our hydrometer corresponded to a density of 1.04 g/mL.

4b. Ten mL of our unsaturated salt solution was measured out in a small graduate and found to have a mass of 10.5 grams: Density = 10.5 g / 10 mL = 1.05 g/mL

    This closely matches the density of 1.04 g/mL read on the hydrometer scale.

    The calculated density of 1.05 g/mL is more reliable, because the mass and volume data on which this figure is based are known with greater accuracy. The hydrometer calibrations are less certain. Small errors in determining the precise water line for all three liquids creates considerable uncertainty on a closely-spaced scale. *(Water adheres somewhat to the plastic straw, creating a range of uncertainty of perhaps 1 mm. Adhesion is strong enough to support the straw at both the high and low end of this range.)*

Materials

☐ Thread.
☐ The improvised hydrometer from the previous activity.
☐ Hydrometer Scales. Photocopy these from the supplementary page at the back of this book. Variations your students are likely to produce will usually fall somewhere within this range of 35 scales. In the unlikely event that your particular hydrometers fall outside this range, photocopy this sheet at 120% to range upward, or reduce it to 80% for narrower scales. Have students share sheets to conserve paper.

☐ A 100 mL graduate.
☐ A water tub.
☐ A 10 mL graduate.
☐ Saturated salt water.
☐ A tablespoon of salt.
☐ A glass of water.
☐ A gram balance.

(TO) observe and explain the interaction of mixing liquids with different densities.

THIN BLUE LINE   O   Floating and Sinking (   )

1. Fill a glass brimful of cold water. Catch overflow in a sink or tub.

   a. Cover with plastic wrap. Be careful not to trap large air bubbles. (A few small ones are OK.)

   b. Seal the water in (and the air out) with a rubber band.

   c. Poke a very tiny hole in middle of the plastic with the very tip of a pin. (Don't push it past its point.)

2. Place several drops of salt water on the pinhole and observe. Follow with a drop of blue food coloring. Alternate between these two liquids.

   a. List 4 or more unique observations. Include drawings where possible.

   b. Explain 2 or more of your observations in terms of density and displacement.

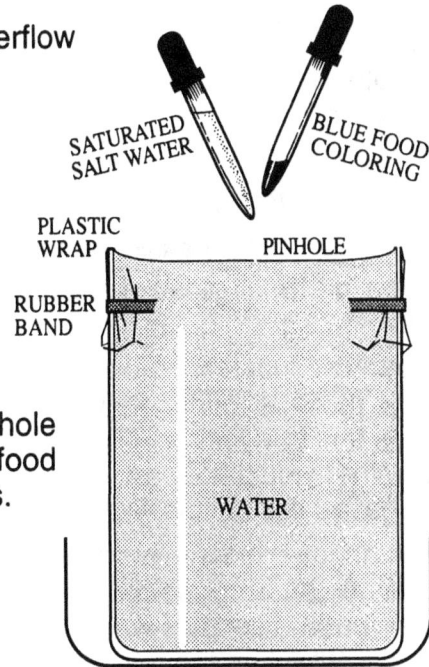

SATURATED SALT WATER

BLUE FOOD COLORING

PLASTIC WRAP

PINHOLE

RUBBER BAND

WATER

© 1995 by TOPS Learning Systems

21

Answers / Notes

2. *There are many things to notice and wonder about. (Is water temperature a variable that matters?) Encourage students to ask what-would-happen-if questions and set about to find the answers.*

• A thin, clear stream of salt water drops rapidly through the pinhole and sinks through the fresh water to the bottom of the jar. *(This is visible because light refracts as it bends through the higher-density stream.)*

   The higher-density salt solution displaces the lower-density fresh water.

• When food coloring is added, the thin, clear stream inside the jar turns blue and sinks more slowly.

   Food coloring has lower density than salt water but higher density than fresh water. Thus, it sinks through fresh water more slowly than salt water does.

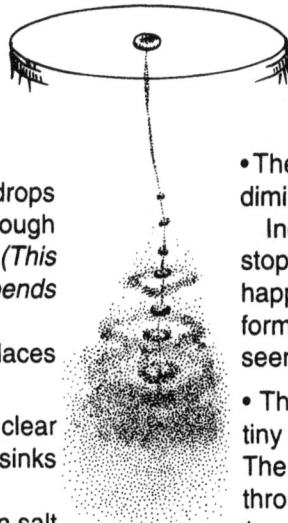

• Water at the bottom of the jar turns blue, while the water higher up remains clear.

   The blue salt water has higher density. It settles to the bottom of the jar, below the clear water.

• The stream volume is relatively large at first, then diminishes to a thin blue line over time.

   Increasing pressure in the jar seems to slow, stop and finally reverse the flow of water. As this happens, a lighter circle of outflowing clear water forms over the pinhole. At other times, the pinhole seems to plug up, perhaps with crystallized salt.

• The continuous stream breaks into a series of tiny expanding rings that descend more slowly. The effect is rings, flowing through rings, flowing through rings. *(In more rapid flows the stream breaks into branches and forms a more complex pattern of rings.)*

• Giving the jar a sudden nudge causes the stream to twist and bend.

Materials

☐ A glass or jar with smooth sides for undistorted viewing.
☐ Plastic wrap.
☐ A rubber band. A pint or quart jar with screw-on canning ring provides an easy-to-seal alternative.

☐ A sink or water tub.
☐ A pin.
☐ Saturated salt water with a dispensing dropper.
☐ Blue food coloring with a dispensing dropper.

(TO) observe that ice, unlike candle wax, is less dense than its own liquid. To appreciate that our environment would be radically changed if ice sank in water.

MELTING AND FREEZING   ⚪   Floating and Sinking (   )

1. Gently melt a few chips of wax in a tablespoon over a candle. Blow out the candle as soon as the last bit of wax melts.

> **CAUTION:** Don't overheat.
> Hot wax can blister your skin.

2. Drop in a small piece of solid wax. Does wax float or sink in its own liquid?

3. Nearly all substances lose density when they melt.
   a. Is this true for candle wax?
   b. Is this true for water? How do you know?
   c. What would happen to lakes on Earth if water behaves like candle wax?

4. Put an ice cube into a small glass half filled with corn oil.
   a. Draw an accurate diagram showing how ice rests in the corn oil.
   b. What happens as water melts off the cube? Show this in your drawing.
   c. Comment on the relative densities of ice, corn oil and water.

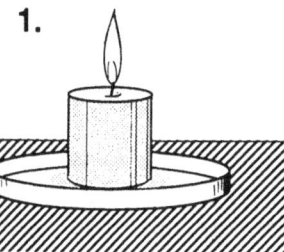

© 1995 by TOPS Learning Systems   22

Answers / Notes

1. *Supervise this step closely. Most students have put fingers into the soft wax of a just-extinguished candle without ill effect. Wax in a tablespoon, however, is easily heated to much higher temperatures if the candle is not extinguished as directed.*

2. Small pieces of solid wax sink to the bottom of the tablespoon, below the surface of the liquid wax. *(If these pieces of solid wax are very tiny, they may be supported at the cooling, solidifying surface. Push them under with a match stick, and they stay submerged.)*

3a. Yes. Solid candle wax must lose density when it melts, because it sinks in its own liquid.

3b. No. Ice in a cold drink always floats on the surface

3c. If frozen water behaved like candle wax, ice would sink to the bottom of winter lakes as soon as it formed on the surface. Then more ice would freeze on the surface and sink as well. Over time, layer upon layer of ice would build up until the *entire* lake froze solid. In areas with moderate summer temperatures, the sun might thaw surface water, but not the entire lake. Fortunately for fish, ice doesn't sink in water. Only surface water freezes during winter, and remelts in the summer sun.

4a,b.

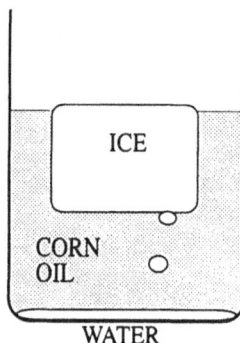

4b. Melted water drips off the bottom of the cube and sinks to the bottom of the corn oil.

4c. Ice floats in corn oil, but water sinks below it. Thus water is most dense, corn oil is less dense, and ice is least dense.

Materials

☐ Small chunks of candle wax.
☐ A candle with a drip catcher and matches.
☐ A tablespoon.
☐ An ice cube.
☐ A small glass half full of corn oil. Or, you can use the small dispensing jar from earlier activities. Corn oil won't be used again in this module.

(TO) ballast a floating candle so it burns just above the water line. To explain its floating and sinking characteristics in terms of density and Archimedes' principle.

FIRE ON WATER ○ Floating and Sinking ( )

1. Stick a pin point straight into the bottom of a birthday candle. Set it in a glass filled *nearly* to the top with room-temperature water.

If it <u>sinks</u>, lift it back out (use straw "chopsticks" if needed.) Nip off bits of the pin with wire cutters until the candle sticks out of the water no higher than it is wide.

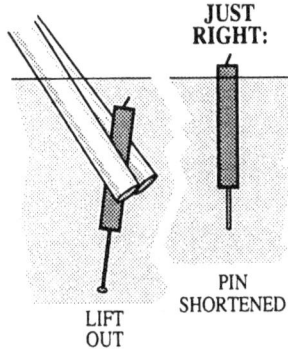

JUST RIGHT:

LIFT OUT

PIN SHORTENED

If it <u>floats</u> too high, lower it with a bit of clay stuck to the pin.

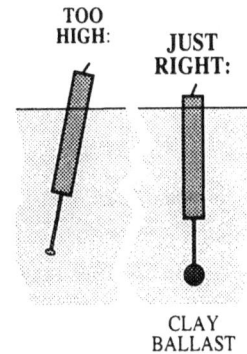

TOO HIGH:

JUST RIGHT:

CLAY BALLAST

2. Predict what happens when you light the candle: will it burn most of the candle, or just a little? Explain your reasoning.

3. Light the candle and test your prediction.
   a. Was it correct? What happened?
   b. Use Archimedes' principle to explain your observations.

© 1995 by TOPS Learning Systems

23

Answers / Notes

1. *The candle and pin may sink or float, depending on the materials used. If students fill the glass brimful (instead of nearly full as directed), the candle will cling to the sides of the glass. When properly weighted, the candle pokes only about 2 mm above the water's surface, and leans very little.*

2. Only a little of the wax will burn. The candle/pin barely floats, with an average density slightly less than the density of water. When you light the candle, the lower-density candle wax begins to burn away, while the higher-density pin remains intact. This will quickly increase its average density, and sink the burning candle. *(This is a well-reasoned prediction, but it is wrong! Few, if any, students will anticipate the interaction of melted candle wax with water.)*

3a. The candle burns for a much longer time than predicted. Melted wax flows out onto the water, forming a solid collar that surrounds the flame like a boat. The flame itself actually sinks below water level as the candle/pin slowly gains average density. After the candle burns to between 1/2 and 1/10 of its original length, the flame grows more and more feeble until it finally goes out. Water never touches the wick, nor does the extinguished candle sink.

3b. The melting wax cools into a "boat" that surrounds and buoys up the burning candle/pin with a force equal to its own weight. As its density increases, so does the displacing capacity of the surrounding boat.

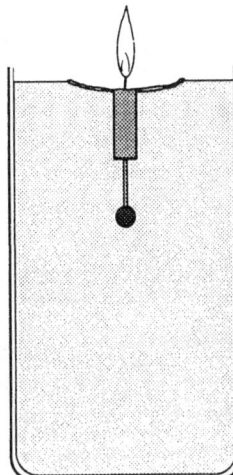

Extension

Is water temperature an important variable in this experiment? (Yes. Heat flows from the candle flame into cold water more rapidly than into hot water, cooling the wax below its combustion temperature much sooner. In hot water the wax spreads out like a raft and burns down completely; in cold water it forms a small modest collar around the candle and burns down only about halfway.)

Materials

☐ A new birthday candle.
☐ A straight pin.
☐ A glass or jar. It must be taller than a birthday candle and pin placed end to end.
☐ Two straws, optional.
☐ Wire cutters.
☐ Clay.
☐ Matches.
☐ A room with calm air.

enrichment

(TO) adjust the weight of a helium-filled balloon to float with neutral buoyancy in room air. To study its floating and sinking properties in warmer and colder air of different densities.

FLOATING IN AIR ○ Floating and Sinking ( )

1. Get a helium-filled balloon. Anchor it to a clay ball on a short leash of string or ribbon.

2. Pinch off bits of the clay until the balloon has neutral buoyancy, neither rising nor sinking much in the still air of your room.

    a. What force supports your balloon so it neither rises nor falls?

    b. A 5 liter (5,000 mL) balloon floats just like yours when enough clay is added to give it a total mass of 6.00 g. What is the density of air surrounding this balloon? Explain your reasoning.

CLAY ANCHOR

3. Float your neutral balloon in air that is colder and/or warmer than the air in your room. What do you observe? What can you conclude?

4. How is your balloon like a thermometer?

24

Answers / Notes

2a. The balloon is buoyed up by a force that is equal to the weight of the room air it displaces.

2b. The balloon floats neutral: it displaces its volume and weight in air. The density of the displaced air is, thus,

$$6.00 \text{ g} / 5 \text{ liters} = 1.2 \text{ g/liter.}$$

3. The neutral balloon tends to rise in colder air, which must have greater density than room air; it buoys up the balloon with greater force. Similarly, the neutral balloon tends to sink in warmer air, which must have less density than room air; it buoys up the balloon with less force, insufficient to support the full weight of the balloon.

4. The balloon gives a relative indication of temperature: the faster it rises, the colder the air temperature; the faster if falls, the warmer the air temperature.

Extension

Make a hot air balloon. Straws, thread, tape and a light plastic garment bag are good construction materials. Don't send your heat source up with the balloon, as it could become a fire hazard. Test the balloon on a cold, calm day.

Materials

☐ A helium-filled balloon with clay anchor. Mold a clay ball around a hand-span length of ribbon or knotted string.
☐ A room with relatively still air. No fans or breezy open windows.
☐ Access to colder and/or warmer bodies of air. Moderate temperature differences (in another room or hallway?) easily demonstrate differences in buoyancy.

REPRODUCIBLE
STUDENT
TASK CARDS

Task Cards Options

Here are 3 management options to consider before you photocopy:

1. Consumable Worksheets: Copy 1 complete set of task card pages. Cut out each card and fix it to a separate sheet of boldly lined paper. Duplicate a class set of each worksheet master you have made, 1 per student. Direct students to follow the task card instructions at the top of each page, then respond to questions in the lined space underneath.

2. Nonconsumable Reference Booklets: Copy and collate the 2-up task card pages in sequence. Make perhaps half as many sets as the students who will use them. Staple each set in the upper left corner, both front and back to prevent the outside pages from working loose. Tell students that these task card booklets are for reference only. They should use them as they would any textbook, responding to questions on their own papers, returning them unmarked and in good shape at the end of the module.

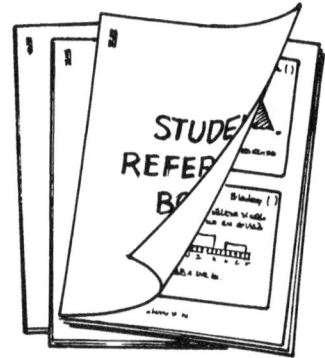

3. Nonconsumable Task Cards: Copy several sets of task card pages. Laminate them, if you wish, for extra durability, then cut out each card to display in your room. You might pin cards to bulletin boards; or punch out the holes and hang them from wall hooks (you can fashion hooks from paper clips and tape these to the wall); or fix cards to cereal boxes with paper fasteners, 4 to a box; or keep cards on designated reference tables. The important thing is to provide enough task card reference points about your classroom to avoid a jam of too many students at any one location. Two or 3 task card sets should accommodate everyone, since different students will use different cards at different times.

DENSITY OF WATER　　○　　Floating and Sinking (　)

1. Get a 10 mL graduated cylinder and a gram balance. Center the balance.

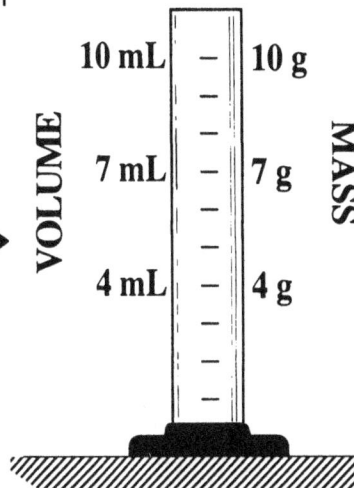

　　a. Show that these *volumes* of water have the indicated *masses*, within the limits of measuring uncertainty:

　　b. Recenter your balance, as necessary, before finding each new mass.

　　c. Write a brief report.

2. *Density* is defined as the mass of any substance divided by its volume. Show that the density of water is always close to 1.00 g/mL, no matter what volume you measure.

$$D = \frac{mass}{volume} = \frac{g}{mL}$$

1

DENSITY OF OTHER LIQUIDS　○　　Floating and Sinking (　)

1. You already found that the density of water is 1.00 g/mL. Now calculate the density of each of these liquids, following the steps below:

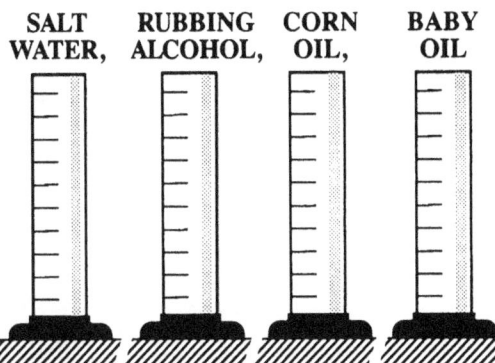

SALT WATER,　RUBBING ALCOHOL,　CORN OIL,　BABY OIL

　　a. Rinse and dry your 10 mL cylinder between each new liquid. Roll and tape a paper towel to reach to the bottom.

　　b. Find the mass of 10.0 mL samples of each liquid to the nearest 0.1 g. Recenter your balance, if necessary, before finding the mass of each new liquid.

　　c. Report all densities to the nearest 0.01 g/mL. Show your math.

2. Order these 4 liquids, plus water, from most dense to least dense.

MOST DENSE　　**LEAST DENSE**

2

LIQUID PAIRS

O   Floating and Sinking (   )

1. There are 6 different ways to pair these 4 liquids: water, rubbing alcohol, baby oil, corn oil. List these pairs on 6 lines of notebook paper.

1. water/rubbing alcohol: one layer
2. _____
3. _____
4. _____
5. _____
6. _____

2. Mix a few drops of each pair, one pair at a time, on the bottom of an inverted jar.

a. If one layer forms, write this next to the liquid pair in your list.

b. If two layers form, write this next to the liquid pair. Further, record which liquid floats on top and which sinks underneath.

c. Wipe the glass dry with a paper towel after each test. Use just 1 towel to conserve trees.

3. How does density determine the floating and sinking behavior of each liquid?

3

PREDICTING LAYERS

O   Floating and Sinking (   )

1. Draw a diagram to *predict* what a test tube will look like if you gently add two droppers full of each liquid in this order:

1st — corn oil
2nd — water
3rd — alcohol
4th — baby oil

a. Explain your reasoning.
b. Test your prediction: tilt the tube so each addition runs down its side.

2. Draw a diagram to *predict* how the test tube will change if you now gently mix the liquids.

LET EACH LIQUID FLOW GENTLY DOWN THE SIDE.

a. Explain your reasoning.
b. Test your prediction: cover the mouth with your thumb and slowly invert the tube *just once.*

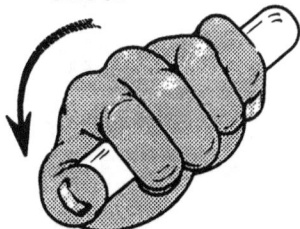

3. Vigorously shake the test tube to mix the liquids.

a. What happens?
b. Do the liquids stay mixed? Explain.

4

DUNK THE CANDLE   O   Floating and Sinking (   )

1. Get a candle with both ends cut square. Find its mass on a centered balance.

2. Bend the head of a pin with pliers to form a hook. Push it into the end cut most evenly. Slide the candle (hook up) into a 100 mL graduate about 1/3 full of water.

PIN HOOK

CANDLE

WATER

    a. Is the density of candle wax greater than 1.00 g/mL, or less? How do you know?

    b. Punch a hole in the end of a straw. Use the hole in this straw to "fish" the candle back out of the water by its hook.

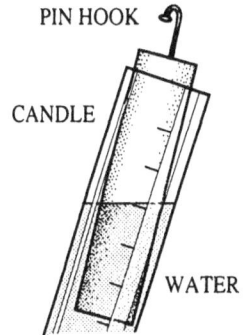

3. Adjust the water level to 40.0 mL, then lower the candle back into the cylinder. Push on the hook with the straw to dunk *all* of the candle under water.

    a. How much water does the candle *displace* (push out of the way)? Find the difference between initial and final water levels.

    b. What is the volume of the candle? Explain your reasoning.

    c. Calculate the density of candle wax in g/mL.

(Save your candle with hook and "fishing" straw to use again.)

© 1995 by TOPS Learning Systems   5

SPECIFIC GRAVITY   O   Floating and Sinking (   )

1. Gently lower your candle by its hook into a 100 mL cylinder full of water so it floats in the hump at the top. Catch overflow in a saucer.

2. Stick a pin into the floating candle precisely where the surface of the water touches its side. Lift the candle out.

3. Drill the pinhole a little larger, and smear a bit of clay in the hole. This marks its floating waterline.

MARK WITH PIN

    a. Cut, roll and tape notebook paper around that part of the candle that floats *under* water.

    b. Label it like this:

CLAY MARK

Displaced Water
(The floating candle pushes away this much water.)

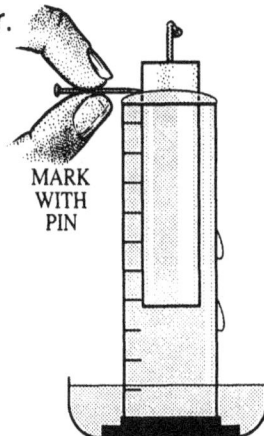

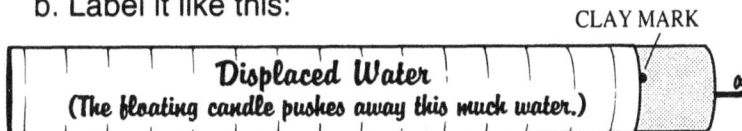

4. Specific Gravity (S.G.) compares the density of any substance to the density of water. Calculate the specific gravity of the candle in two different ways:

    a.   S.G. = $\dfrac{\text{Density of candle}}{\text{Density of water}}$ b.   S.G. = $\dfrac{\text{Length of displaced water}}{\text{Length of candle}}$

© 1995 by TOPS Learning Systems   6

FLOAT THE CANDLE ○ Floating and Sinking (   )

1. Slide the wrapper off your candle. Check that your clay dot still marks the candle's float line in a 100 mL graduate brimful of water.

2. Now fill the graduate with 40.0 mL of water. Gently float the candle, without splashing.

    a. Read the volume where the clay mark floats. (This is the true float line, even if water creeps higher in the narrow space between candle and cylinder wall.)

    b. Find the difference between initial and final volumes. This displaced water equals the volume of your paper cylinder.

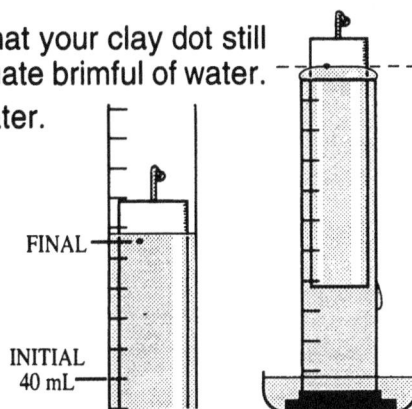

FINAL

INITIAL 40 mL

3. You know that water has a density of 1.00 g/mL. This means that each mL of displaced water has a mass of 1 gram.

    a. What is the total mass of water represented by your paper cylinder?

    b. Recall the total mass of your candle from a previous activity. Compare this value to the mass of its displaced water.

    c. What seems to be the relationship between the mass of a floating candle and the mass of water it displaces?

© 1995 by TOPS Learning Systems

7

HOW NOW, BROWN DOWEL? ○ Floating and Sinking (   )

Get a large wood dowel with both ends cut square.

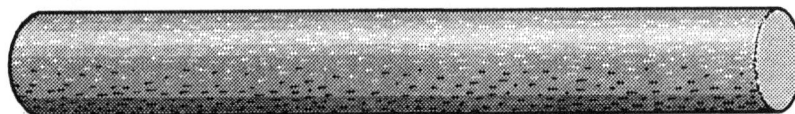

1. Find its density.

2. Find its specific gravity using 2 different ratios.

3. Show that it displaces a mass of water equal to its own mass.

Review these 3 task cards!

DUNK THE CANDLE 5
SPECIFIC GRAVITY 6
FLOAT THE CANDLE 7

© 1995 by TOPS Learning Systems

8

THE NATURAL ORDER   O   Floating and Sinking (   )

1. Roll precisely 50.0 g of clay into a cylinder shape so it fits inside a 100 mL graduate. Stick a bent pin hook in one end that you can catch in the hole of your "fishing" straw as before.

a. Use water displacement to find the volume of this clay.

b. Calculate its density.

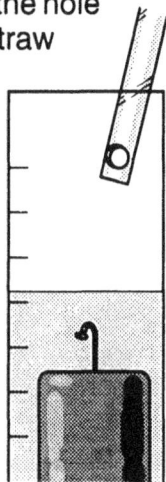

2. You know the densities of 5 liquids and 3 solids. List these substances in order, from least dense to most dense, along with your experimental data.

Least Dense

_____
_____
_____
_____
_____
_____

Most Dense

3. Build up layers of liquids and solids in a test tube or 10 mL graduate. Did everything behave as you expected? Explain.

BIRTHDAY CANDLE

TOOTHPICK

CLAY

9

OVERFLOW CUP   O   Floating and Sinking (   )

1. Make 2 parallel cuts, 1 cm wide, about 1/4 of the way down the side of a paper cup. Fold out the flap to make a spout on an *overflow cup.*

2. Set this cup on an inverted can beside a plastic tub. Fill it with water until it overflows into the tub.

GOLF BALL

3. Replace the tub with a large graduated cylinder. Gently drop a golf ball into the cup and measure all the displaced water that overflows into the graduate.

4. What is the volume of the golf ball in mL? Explain how you know.

5. Set the golf ball between 2 batteries. Measure its diameter (the distance between batteries) with a centimeter ruler. What is its radius?

6. The volume of a ball in cm³ is given by $V = 4/3\pi r^3$. Use a calculator to find this volume.

BALL'S DIAMETER

7. Compare your volumes in steps 4 and 6. What can you say about the relationship between milliliters (liquid measure) and cubic centimeters (dry measure).

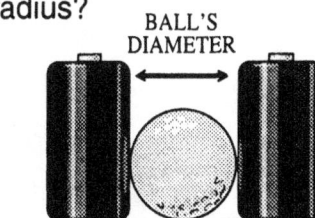

10

DENSITY MATH ○ Floating and Sinking (   )

1. Roll a lump of clay that has about the same volume as a size-D battery.

   a. Pull off about 1/3 of this clay and shape it into a cube. Shape the remaining 2/3 like a pyramid.

   b. You have previously determined the density of this clay. What is it?

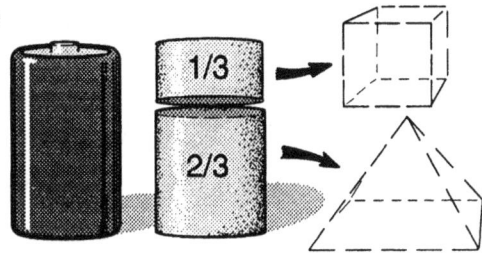

2. Find the volume of the pyramid with your overflow cup.

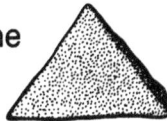

   a. Multiply the pyramid's volume by its density to find its mass.

   volume x density = mass
   mL x g/mL = g

   b. Check this result on your balance.

3. Find the mass of the cube on your balance.

   a. Divide the cube's mass by density to find its volume.

   mass/density = volume
   g x mL/g = mL

   b. Check this result with an overflow cup.

11

CLAY BOAT ○ Floating and Sinking (   )

1. Mold a 50.0 g mass of clay into a flat "pancake."

2. Wrap a size-D battery, *from the bottom up*, in a square of plastic wrap.

   a. Mold your clay around the battery, *from the bottom up*. Extend the sides nearly to the top.

   b. Remove the battery and pull out the plastic.

   c. Hold your clay "boat" up to bright light to detect holes. Fix them so it floats in water with no leaks.

   d. Check that your boat still has a mass of 50.0 g.

3. What volume of water will your clay boat displace in an overflow cup? *First* make a reasoned prediction, then test.

4. What volume of water will this clay displace if you squash it together and roll it into a sphere? *First* make a reasoned prediction, then test.

12

NEITHER HERE NOR THERE ○   Floating and Sinking ( )

1. Cover the cork supplied by your teacher with a thin "pancake" of clay until you can't see the cork.

2. Put this clay/cork mass in a glass of water at room temperature. If it floats, shake off the water and add more clay. If it sinks, scrape clay away.

   a. Make fine adjustments until it *both* remains at the water's surface *and* sinks when you push it below the surface. Smooth the clay evenly around the cork when you find the right amount.

   b. What is the average density of your clay/cork? (No calculations necessary!)

3. Push your clay/cork to the bottom of the glass. It should remain there, not quite rising.

   a. Trickle eye droppers full of saturated salt water down the side of the glass, until the clay/cork neither sinks nor floats.

   b. Explain what's happening.

4. Save your clay/cork in a small plastic bag labeled with your name.

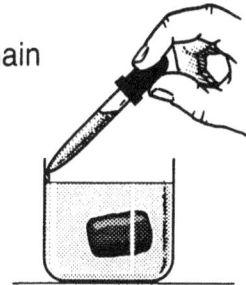

CLAY-COVERED CORK

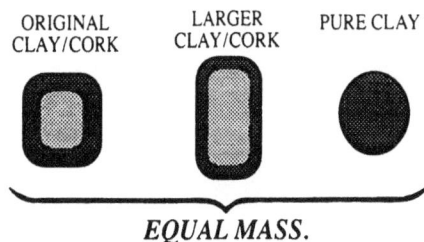

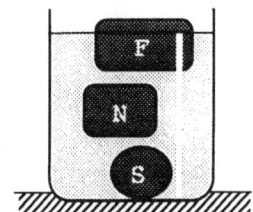

© 1995 by TOPS Learning Systems   13

THREE CLAY BODIES ○   Floating and Sinking ( )

1. Get your clay/cork from its plastic bag. Find its mass on a gram balance.

   a. Get another cork, about twice as large as the first. Cover it with clay so it has precisely the same mass as your original clay/cork.

   b. Roll a ball of pure clay that has precisely the same mass as your large and small clay/corks.

ORIGINAL CLAY/CORK   LARGER CLAY/CORK   PURE CLAY

EQUAL MASS.

2. Put each clay body in water. Lightly press an F, N, or S into each one with your pencil to indicate whether it Floats, is Neutral, or Sinks in water.

3. Which of these 3 bodies displaces its own *mass* in water? Use your overflow cup and a 100 mL graduate to evaluate.

4. Which of these 3 bodies displaces its own *volume* in water? Explain your reasoning.

5. F, N, and S all have the same mass. Why, then, do they behave differently in water?

6. Store all 3 clay bodies in your labeled plastic bag.

© 1995 by TOPS Learning Systems   14

SPRING SCALE ◯ Floating and Sinking (   )

1. Cut out a Weight Scale. Tape 4 short lengths of straw to it where shown.

2. Loop a rubber band near the side of a big cereal box. Fold your Weight Scale around the middle of that side, and tuck the straws under the rubber band.

3. Cut 1.5 meters of steel wire. Form a *very* tight knot 11 cm from one end. Coil it closely and evenly (without overlap) around your pencil, starting 5 cm beyond the knot.

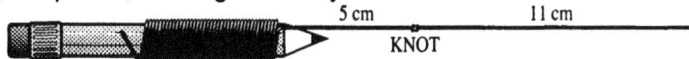

4. Slide this spring off your pencil. Straighten the last coil. Tape it to the top of your box so the spring hangs over the scale.

5. Pull out the "arm" of a paper clip a little, to form a hook. Wire it to the bottom of your spring.

6. Tape a 15 cm thread "bucket handle" to a film canister. Hang it on the hook, over your table edge.

7. Stretch out the coils of your spring so they hang halfway down the box *when relaxed.* (Pull them a little past halfway, then release.)

8. Slide the scale down and out until the "0" line touches the knot. This centers your spring scale.

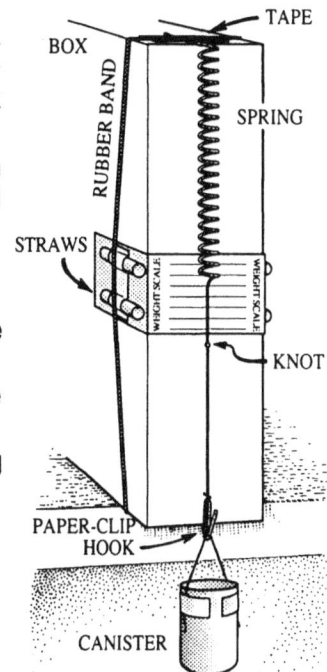

15

FRESH WATER WEIGH-INS ◯ Floating and Sinking (   )

1. Stick a pin through the sticky side of a masking tape square. Tape it on the bottom of the canister on your spring scale, point down.

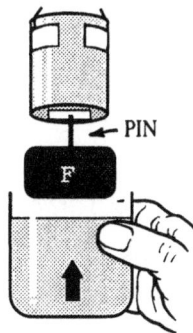

2. Center the knot at "0" weight units by adjusting the movable scale

3. Stick clay body F on the pin.
   a. Weigh it in air; in a glass of water.
   b. Copy this table. Write your results in the top row of the first 2 columns.

4. Recenter your balance. Record weights for N and S in air and in water.

5. An object immersed in water is pushed up by a force called *Buoyancy*. Calculate the buoyancy of water on each body and fill in the third column.

$$B = (Wt)_{Air} - (Wt)_{Water}$$

FRESH WATER TABLE (Weight Units)

	Weight in AIR	Weight in WATER	BUOYANCY of Water	Weight Water DISPLACED
Body F				
Body N				
Body S				

6. Collect the water displaced by each body in your overflow cup, and weigh it. Fill in the fourth column.

7. Within the limits of experimental error, what generalizations can you make?

16

SALT WATER WEIGH-INS ○   Floating and Sinking ( )

1. Copy and complete this table for *salt* water as you did for fresh water.

SALT WATER TABLE (Weight Units)

	Weight in AIR	Weight in SALT WATER	BUOYANCY of Salt Water	Wt Salt Water DISPLACED
F				
S				

2. Read Archimedes' principle. Does it hold for salt water? Explain.

> ### Archimedes' Principle
> An immersed object is buoyed up by a force equal to the weight of the fluid it displaces.

3. Compare *fresh* water data (from before) and this salt water data for the <u>S</u>inking body:

   a. Are equal weights of both liquids displaced? Why?

   b. Does S weigh the same in both liquids? Why?

FRESH WATER   SALT WATER

4. Compare *fresh* water and *salt* water data for the <u>F</u>loating body:

   a. Does F weigh the same in both liquids? Why?

   b. Would you expect the volumes of each displaced liquid to be equal? Predict, then test.

17

ARCHIMEDES' PRINCIPLE ○   Floating and Sinking ( )

1. Fill the canister on your spring scale brimful of water. Put a tub underneath to catch the overflow.

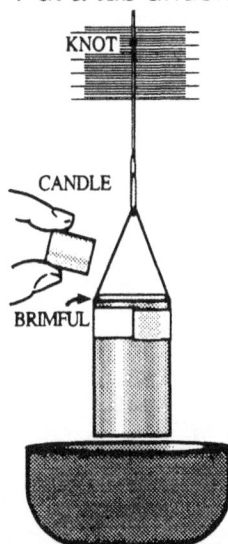

KNOT

CANDLE

BRIMFUL

   a. If you float a piece of a candle in the canister, will the knot move lower? Use Archimedes' principle to support your answer.

   b. Test your prediction.

2. Dry the canister and recenter your spring scale to "0".

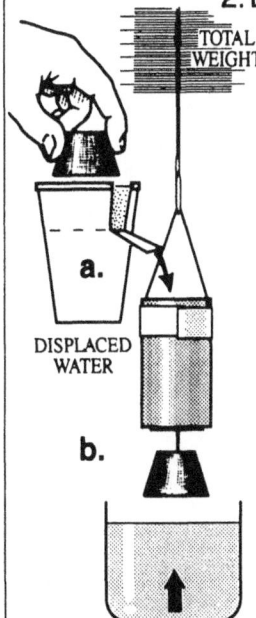

TOTAL WEIGHT

a. DISPLACED WATER

b.

   a. Drop a rubber stopper into an overflow cup brimful of water. Catch all water it displaces in your canister.

   b. Hang the stopper plus displaced water on your scale. Buoy up the stopper in a glass of water, and read the total weight on your spring scale.

   c. Does this weight equal the stopper's weight? Use Archimedes' principle to support your answer.

   d. Test your prediction.

18

HYDROMETER (1)   O   Floating and Sinking ( )

1. Close one end of a straw with a small plug of clay.

2. Melt a puddle of wax onto a plastic lid from a lighted candle (about a dozen drops). When the liquid starts to solidify, dip in the plugged end to seal with wax.

3. Set a 100 mL graduate, filled to the top with water, in an overflow saucer. Add BB's to your plugged straw so only 1/5 of it floats above water.

4. Close the straw with masking tape. Trim off excess tape.

5. Your straw should now float in a "hump" of water above the graduate's mouth. If not, add more water.

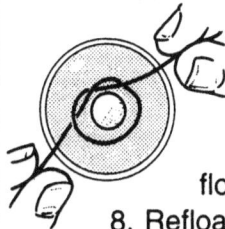

6. Lower a dry loop of thread over the straw until it floats on this hump. Pull it tightly around the straw to mark the float line.

7. Lift the straw from the water. Mark the float line with pencil, then remove the thread.

8. Refloat the straw in the hump of water to check your float line for accuracy. If necessary, erase and remark.

9. You have just made a hydrometer, an instrument that measures density. How do you think it works?

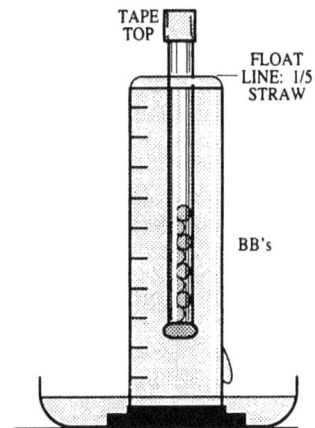

TAPE TOP

FLOAT LINE: 1/5 STRAW

BB's

19

HYDROMETER (2)   O   Floating and Sinking ( )

1. You already marked the float line for water on your hydrometer at 1.00 g/mL.
   a. Recall the density of saturated salt water. Mark a new float line for this liquid by repeating steps 5-7 from the last activity.
   b. Refloat your hydrometer in a "hump" of salt water to check that this new float line is accurately placed. If not, erase and remark.

2. Cut a rectangle of paper with an edge equal to the distance between both density values on your hydrometer. Label it like this:

Fresh Water = 1.00 g/mL

Salt Water = (your value)

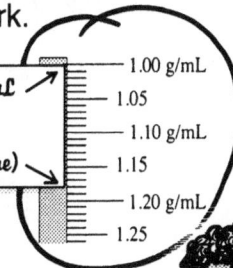

1.00 g/mL
1.05
1.10 g/mL
1.15
1.20 g/mL
1.25

3. Get a sheet of Hydrometer Scales. Find the scale that best fits this rectangle when matched against your pair of density values. Circle it.

4. Dissolve a tablespoon of salt in a glass of water.
   a. Measure the density of this solution with your hydrometer and scale. Tell how you did this.
   b. Find its density by another method and compare values. Which method do you think is most reliable, a or b? Explain.

20

THIN BLUE LINE

Floating and Sinking ( )

1. Fill a glass brimful of cold water. Catch overflow in a sink or tub.

   a. Cover with plastic wrap. Be careful not to trap large air bubbles. (A few small ones are OK.)

   b. Seal the water in (and the air out) with a rubber band.

   c. Poke a very tiny hole in middle of the plastic with the very tip of a pin. (Don't push it past its point.)

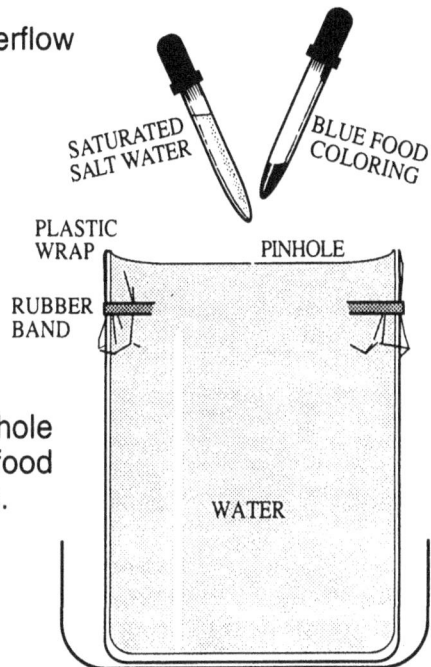

2. Place several drops of salt water on the pinhole and observe. Follow with a drop of blue food coloring. Alternate between these two liquids.

   a. List 4 or more unique observations. Include drawings where possible.

   b. Explain 2 or more of your observations in terms of density and displacement.

SATURATED SALT WATER   BLUE FOOD COLORING

PLASTIC WRAP   PINHOLE

RUBBER BAND

WATER

21

MELTING AND FREEZING

Floating and Sinking ( )

1. Gently melt a few chips of wax in a tablespoon over a candle. Blow out the candle as soon as the last bit of wax melts.

> **CAUTION:** Don't overheat. Hot wax can blister your skin.

2.

1.

2. Drop in a small piece of solid wax. Does wax float or sink in its own liquid?

3. Nearly all substances lose density when they melt.

   a. Is this true for candle wax?

   b. Is this true for water? How do you know?

   c. What would happen to lakes on Earth if water behaved like candle wax?

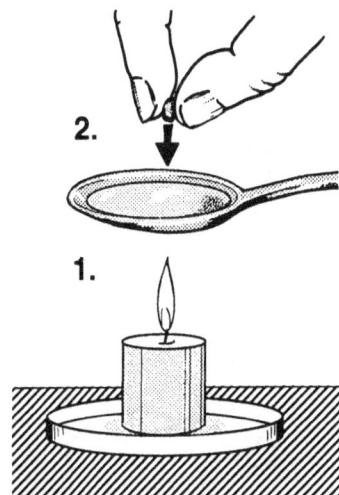

4. Put an ice cube into a small glass half filled with corn oil.

   a. Draw an accurate diagram showing how ice rests in the corn oil.

   b. What happens as water melts off the cube? Show this in your drawing.

   c. Comment on the relative densities of ice, corn oil and water.

22

FIRE ON WATER ○ Floating and Sinking (   )

1. Stick a pin point straight into the bottom of a birthday candle. Set it in a glass filled *nearly* to the top with room-temperature water.

If it <u>sinks</u>, lift it back out (use straw "chopsticks" if needed.) Nip off bits of the pin with wire cutters until the candle sticks out of the water no higher than it is wide.

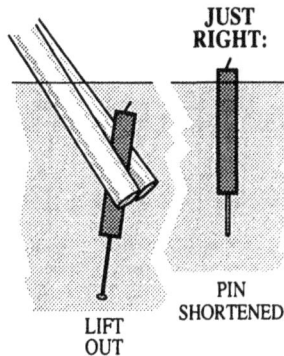

JUST RIGHT:

LIFT OUT

PIN SHORTENED

If it <u>floats</u> too high, lower it with a bit of clay stuck to the pin.

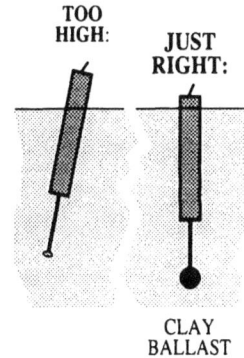

TOO HIGH:

JUST RIGHT:

CLAY BALLAST

2. Predict what happens when you light the candle: will it burn most of the candle, or just a little? Explain your reasoning.

3. Light the candle and test your prediction.
   a. Was it correct? What happened?
   b. Use Archimedes' principle to explain your observations.

23

FLOATING IN AIR ○ Floating and Sinking (   )

1. Get a helium-filled balloon. Anchor it to a clay ball on a short leash of string or ribbon.

2. Pinch off bits of the clay until the balloon has neutral buoyancy, neither rising nor sinking much in the still air of your room.
   a. What force supports your balloon so it neither rises nor falls?
   b. A 5 liter (5,000 mL) balloon floats just like yours when enough clay is added to give it a total mass of 6.00 g. What is the density of air surrounding this balloon? Explain your reasoning.

CLAY ANCHOR

3. Float your neutral balloon in air that is colder and/or warmer than the air in your room. What do you observe? What can you conclude?

4. How is your balloon like a thermometer?

24

SUPPLEMENTARY
CUTOUTS

ACTIVITY 15
WEIGHT SCALE

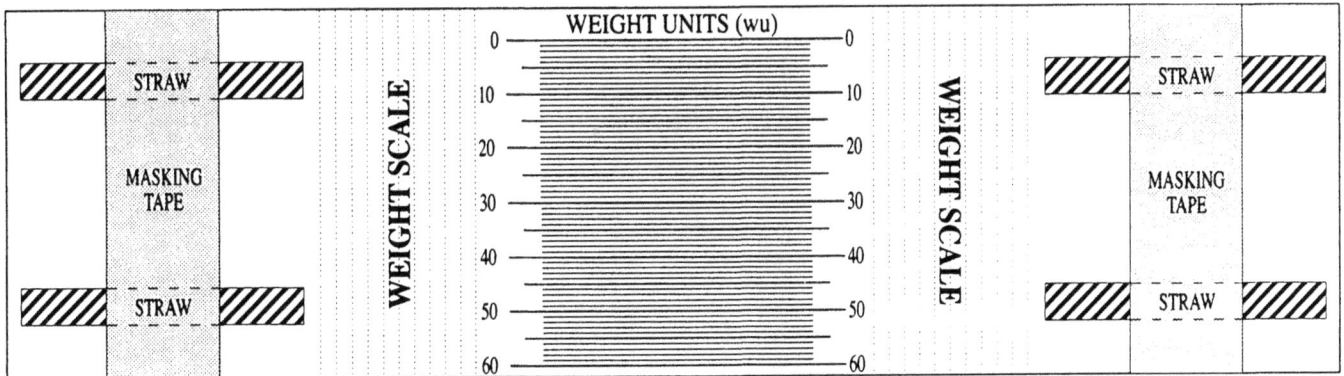

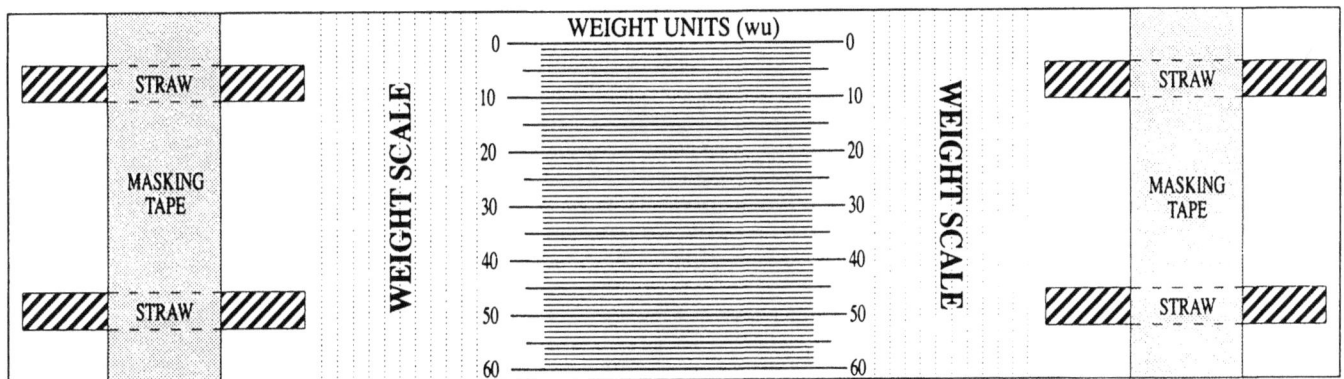

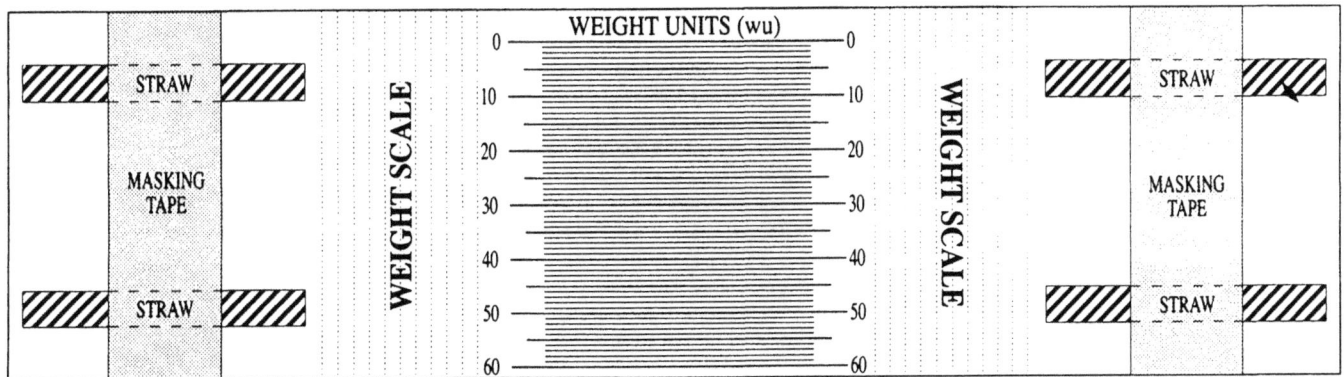

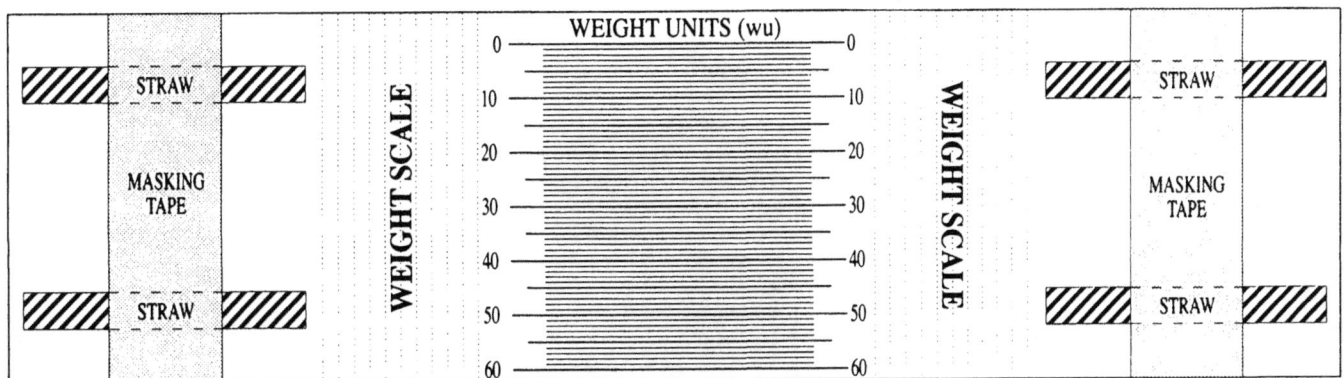

ACTIVITY 20
HYDROMETER SCALES

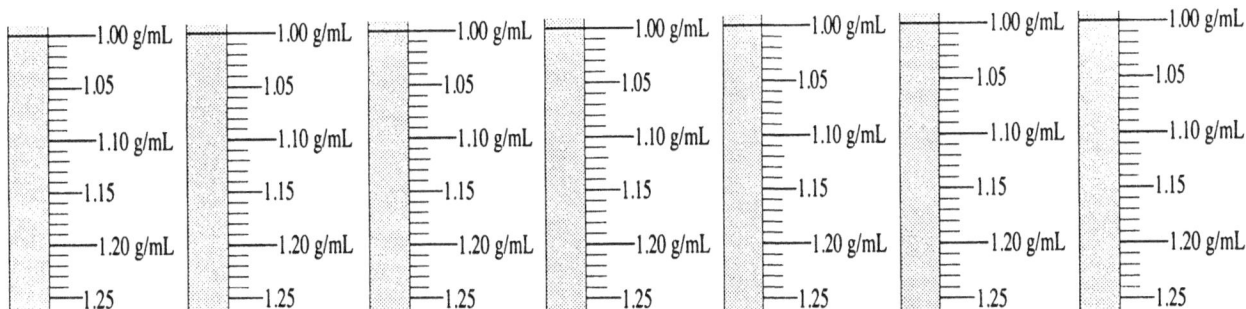

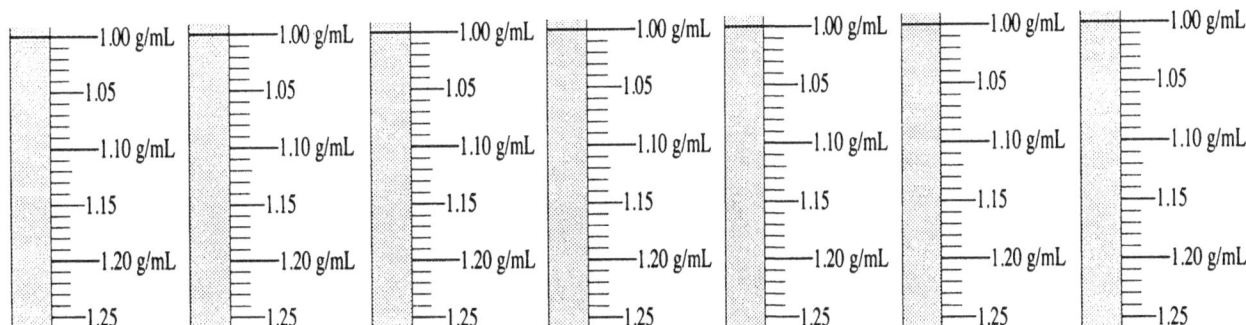

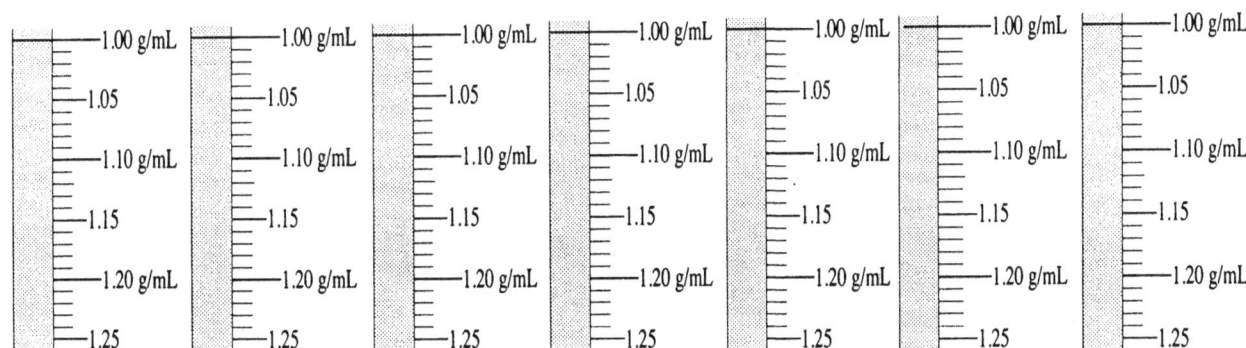

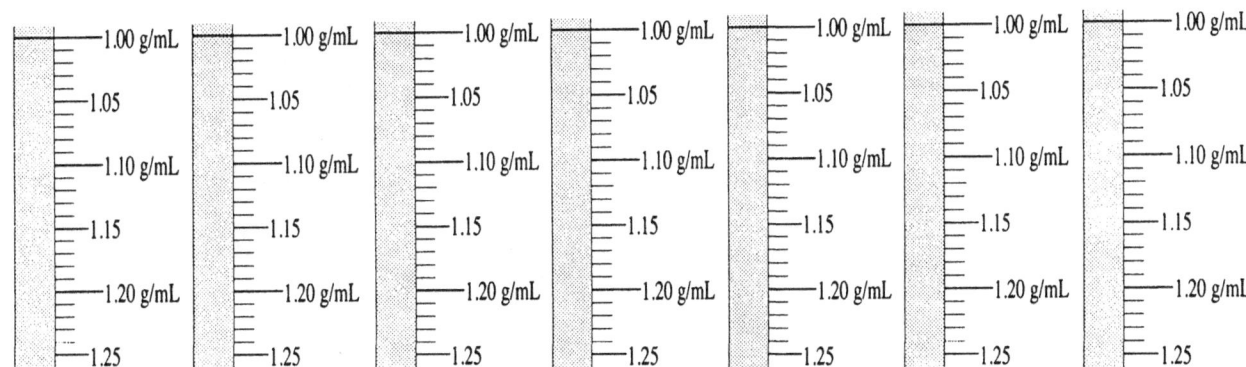

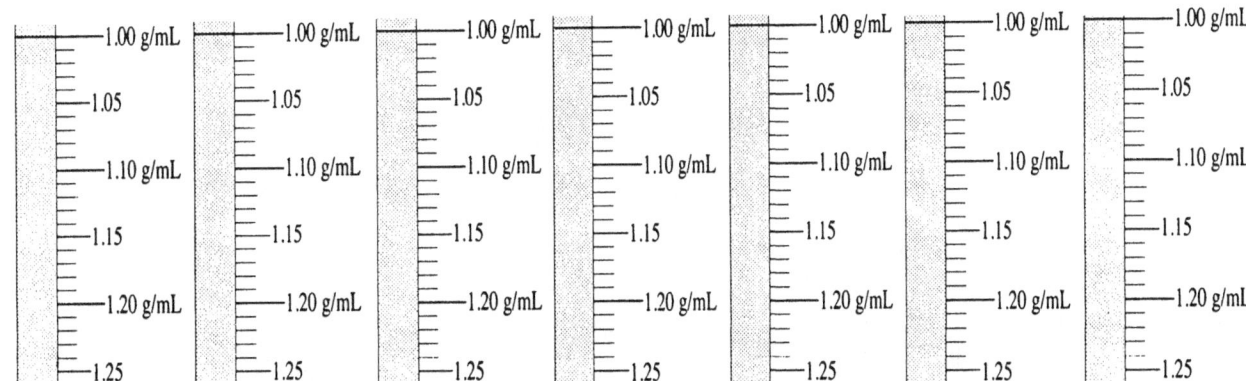

Feedback

If you enjoyed teaching TOPS please tell us so. Your praise motivates us to work hard. If you found an error or can suggest ways to improve this module, we need to hear about that too. Your criticism will help us improve our next new edition. Would you like information about our other publications? Ask us to send you our latest catalog free of charge.

For whatever reason, we'd love to hear from you. We include this self-mailer for your convenience.

Sincerely,

Ron and Peg Marson
author and illustrator

Your Message Here:

Module Title _____ Date _____

Name _____ School _____

Address _____

City _____ State _____ Zip _____

―――――――――――――――――――――――――――― FIRST FOLD ――――――――――――――――――――――――――――

―――――――――――――――――――――――――――― SECOND FOLD ――――――――――――――――――――――――――――

RETURN ADDRESS

_____

_____

_____

PLACE
STAMP
HERE

TOPS Learning Systems
342 S Plumas St
Willows, CA 95988

TAPE HERE

www.ingramcontent.com/pod-product-compliance
Lightning Source LLC
Chambersburg PA
CBHW081513200326
41518CB00015B/2484

9 780941 008792